HALLEY'S COMET *FINDER*

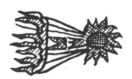

BEN MAYER, A.R.C.A.

A PERIGEE BOOK

Perigee Books
are published by
The Putnam Publishing Group
200 Madison Avenue
New York, NY 10016

Front cover photo by Ben Mayer
Back cover illustration by Ben Mayer
Book design by Ben Mayer and Arlene Schleifer Goldberg
Illustrations, star maps, and astrophotography by Ben Mayer

Library of Congress Cataloging in Publication Data

Mayer, Ben.
  Halley's comet finder.

  Bibliography: p.
  1. Halley's comet. 2. Astronomy—Amateurs' manuals.
I. Title.
QB723.H2M35   1985   523.6   85-9306
ISBN 0-399-51137-7

Starframes and Problicom are trademarks of the author.

Printed in the United States of America
  2 3 4 5 6 7 8 9 10

The author is grateful to Stephen J. Edberg, coordinator for amateur observations of the International Halley Watch at the Jet Propulsion Laboratory, who proofread the manuscript. He communicates the importance of amateur participation in this historic endeavor. Thanks are also due to Dennis di Cicco for permission to reproduce his photograph of Comet Kohoutek, and to the Jet Propulsion Laboratory at the California Institute of Technology for permission to reproduce the photograph of Halley's Comet as it was found on October 16, 1982.

Contents

1. Eureka, There's the Comet!

It depends on you. If you plan to see Halley's Comet you will surely see it. The greater your interest, the *sooner* you will see it. Even now Halley's Comet is moving toward the sun at spacecraft speeds of thousands of miles per hour. Its predicted coming was visually confirmed from Mount Palomar in California in October 1982. Delicate photosensors caught first sight of the interplanetary wanderer through traces of reflected sunbeams that were as faint as the light of a candle seen from a distance of 27,000 miles.

The 1985–86 apparition will favor observers in the Southern Hemisphere of our earth, those who dwell below the equator in South America, Africa and Australia. But skygazers in the United States and other northern regions need not feel left out. Three periods will present themselves for firsthand viewing from these higher latitudes, starting in November 1985 through the end of January 1986, then again from early March into April, and finally in the summer of 1986.

With just a little preparation you can experience the thrill of making your own comet discovery, share the genuine excitement of a scientific search, and have an unforgettable EUREKA experience.

Do not expect a spectacular first moment of splendor at the sight of the comet. Your first sighting will not be of a moon-bright disc with a dazzling tail stretching across the heavens from the eastern to the western horizon. Comets start out far beyond the reach of even the most powerful telescopes, and your first glimpse may yield only the faint approaching nucleus. Tails do not develop until comets move closer to the sun. But when the comet is still far away you can practice finding the slowly emerging promised target and acquaint yourself with simple methods for earliest detection. Soon you can follow the comet's passage during autumn nights, winter evenings and predawn mornings, back into spring nights.

The 1985–86 apparition will be the first where Halley's Comet can be viewed with television cameras. The video arts were not even invented in 1910 when the visitor last came by. No doubt, the comet will be shown on TV in images taken by major telescopes and from high-flying spacecraft, recorded with the latest equipment. There will be pictures showing detail far beyond what earthbound telescopes could ever see. Studying these close-up pictures carefully can and will shed light on your own firsthand observations.

But beware of the news media: Comet Kohoutek was billed as the "comet of the century" in 1973, yet it did not present as big a display as had been predicted and became the astronomical disappointment of the seventies. When, barely eighteen months later, Comet West was approaching, soon to make glamorous naked-eye predawn appearances for days on end, only a few people were aware of the spectacle. The media barely made mention of the grand display. They had become "comet shy" and stayed away.

Years from now, it will be your own riveting first look at the faint glimmer of the comet that you will always remember. You will treasure that moment when suddenly you knew you were observing the same comet that Julius Caesar, Queen Elizabeth I and Isaac Newton viewed in their lifetimes and which history has recorded since before the birth of Christ. Observing an object that has for over two thousand years returned to earth's vicinity at regular seventy-six-year intervals can evoke an awareness of unity with skygazers long gone. Witnessing this taper in the sky will unite starlovers in

modern California with shepherds in ancient Judea who saw the comet over Jerusalem before the birth of Christ.

The 1910 Halley passage also had its firsts. Never before that time had the comet been pictured in photographs. Even in 1910, only a handful of major observatories that had camera equipment could record the comet and its tail with black-and-white photographs.

Today, anyone with a simple 35mm. camera can capture the historic event in color and keep meaningful records to study the progress of the comet from night to night. Your photographs can also be shared in the future with those who have missed seeing this rare happening for themselves.

Even though scientific enlightenment has replaced the superstitious fears of war, death and destruction that beset medieval cometwatchers at the sight of a "bearded star," it may be wise to contemplate the world we will leave to

Edmund Halley, 1656–1742.

Old Royal Observatory, Greenwich.

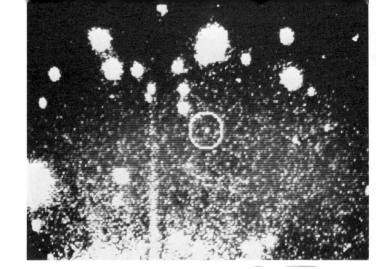

Rediscovery photograph of Halley's Comet: The comet (circled) was found by Caltech astronomers on October 16, 1982. Photographed with the 200-inch Hale telescope on Mount Palomar, California. (California Institute of Technology)

those who may watch the comet during its next appearance in the year 2061. Can we anticipate with hope and confidence a peaceful earth from which future generations of celestial spectators will observe apparitions to come?

As for the immediate tomorrow, the nucleus of another comet, soon to become more dazzling and spectacular than Halley's, may at this very moment be heading our way in an unknown, yet-to-be-computed orbit. It could even outshine the venerable old-timer in dazzle, but not in historic glory. In fact, in 1910 there was such an unexpected companion. If you study and practice the simple disciplines listed in this guide, you may join the ranks of the comet seekers who have discovered such unexpected comets.

During almost any night of the year you may spot such roamers if you know when and where to look. As many as a dozen may be within sight over the course of a year, and many comets are now discovered by amateurs. Whereas once a thorough knowledge of the sky combined with hours at the eyepiece of a telescope were a prerequisite to discovery, today anyone with a camera can join in the comet quest. If you discover a comet it will be named after you, etching your name into celestial annals for all time. No experience is required. Halley's Comet affords a wonderful test object. Let it be your first comet "discovery."

The 200-inch Hale telescope at the Mount Palomar observatory.

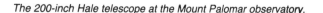

2. Tracking the Comet Through Time and Space

Strangely enough, Edmund Halley (rhymes with *alley*) did not discover the comet that bears his name. His claim to fame is that he was the first to figure out that comets seen in the years 1531, 1607 and, during his lifetime, in 1682 were one and the same. Using Isaac Newton's new mathematics, Halley computed the orbit for this object and dared to predict the periodic returns for the future. Thus he foretold a reappearance for the year 1758. When the comet was sighted again twenty-six years after Halley's death, it was named to commemorate the famous English astronomer and mathematician.

In touching words, blending pride in heritage with an appeal for accuracy, Halley wrote in 1705: "Wherefore if it [the comet] should return according to our prediction about the year 1758, impartial posterity will not refuse to acknowledge that this was first discovered by an Englishman."

Since that time the comet has reappeared in 1835 and in 1910. We are about to witness the 1985–86 appearance. Our children and grandchildren may see it again in the year 2061, and the comet should continue to return for centuries to come.

Chronicles document a two-thousand-year history of past Halley visits. Like a space–time machine, its visitations to us at four-generation intervals are once-in-a-lifetime happenings, but never before could so many be alerted to such an adventure and be shown when, where and even how to witness the historic encounters for themselves. In catching even an initial glimpse of the yet distant comet, you may be looking at the very stuff of which our solar system was created. Later, when the gossamer tail unfolds, you can actually observe the tenuous pressure of sunlight as it pushes cometary matter across millions of miles of space. Even if the 1986 swing of the comet around the sun places it in less than perfect position for easy viewing, it is still the same giant that in 1222, a mere ten passes ago, was so favorably positioned that it could be seen over China in bright daylight.

As we reflect on Halley's history, let's not forget that this messenger from the past is also an emissary to the future. In the year 2366, five appearances from now, Halley may pass so auspiciously that its reflected sunlight will cast night shadows on translunar space habitats.

Any observations that we make in connection with Halley's Comet will become invaluable for future starwatching. The earlier you spot it for yourself, the more exciting the following months will be. Your first observation will be an unforgettable experience, whether you capture the arrival on film or with your eyes. If you follow the comet, it will lead you on a merry chase, starting in November 1985, through the constellations of Taurus, Aries, Pisces and on into Aquarius. There you will lose it in the January evening twilight, only to recover it in the darkening late-February morning skies—still in the Zodiac, but it will now be in Capricorn, gradually growing a tail and heading for Sagittarius. Here Northern Hemisphere observers will have to relinquish custody of their celestial visitor for a while to afford our Southern Hemisphere neighbors their own enviable display before the comet becomes visible again in mid-April. Then we can once more try to be the first to spot our comet in order to pursue it on its outward journey, until it will be lost in the glare of the sun in August 1986.

During these last months of the visit, under warmer spring and summer night skies, we can perform our most useful work for the future. We will be able to sharpen our observing skills on the comet as it gradually fades from view. This will prepare us for exciting events to come—for viewing new comets, because arriving and departing comets look very much alike. Here we can perfect our techniques for comet searching and for blinking (see Chapter 11), which can put you on the threshold of astronomical discoveries. A simple camera and a home-built STEBLICOM discovery device (see Chapter 13) are all you need to start.

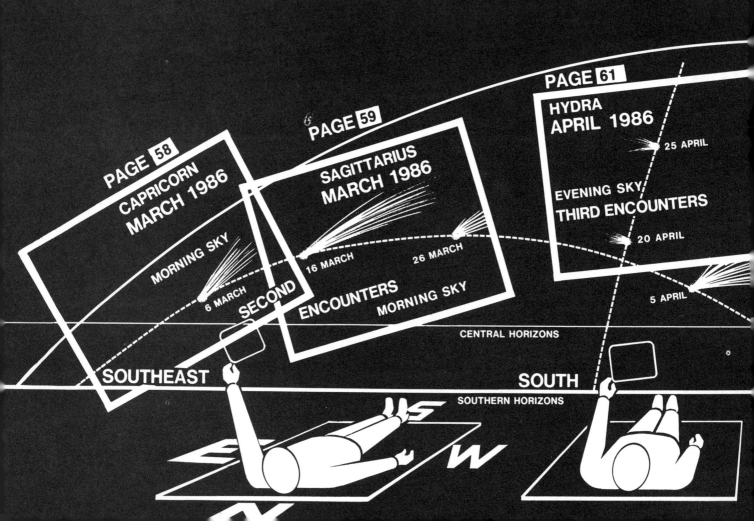

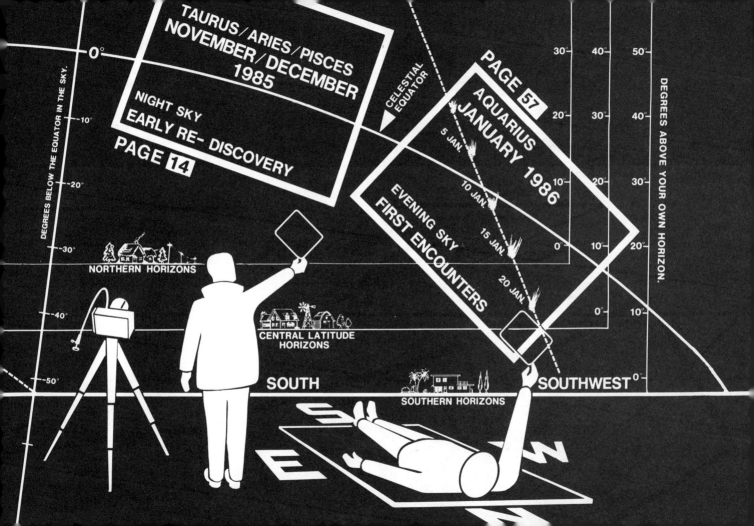

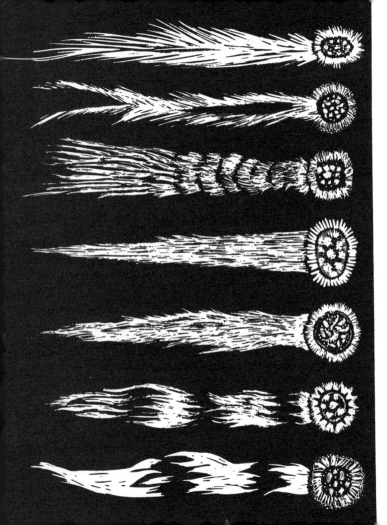

3. Hairy Stars or Methane Sundaes?

In the time of Aristotle, the famous Greek philosopher (384–322 B.C.), and for nearly two thousand years thereafter, comets were believed to be atmospheric "fiery exhalations" appearing unexpectedly at heights somewhere between the earth and the moon. The word "comet" stems from the Greek; *kometes asteres* were "hairy stars." Thus the strange visitors in the sky were also known as "bearded stars." In time the shapes of comet tails reminded the fearful and the superstitious of swords or scimitars, symbols of death and destruction.

Tycho Brahe, a Danish astronomer who lived from 1546 to 1601, fifty years before telescopes were invented, was among the first to make serious scientific investigations of comets. All his research was performed with the naked eye. He built large instruments that allowed accurate measurements to be taken of the positions and the motions of objects in the sky.

While studying observations of a bright comet (not Halley's) made from widely separated locations in 1577, Brahe was able to establish that comets move through interplanetary space far beyond the orbit of the moon, and not through the "upper air" of our atmosphere as Aristotle had stated.

Comets were thought by some to travel in straight lines, and the brilliant German astronomer Johannes Kepler supported this erroneous belief with incorrect computations for the comet of 1607 (Halley's Comet during an earlier visitation). Kepler was an assistant to Tycho Brahe and later achieved great fame for establishing laws governing the elliptical orbits of planets in the solar system.

It was Isaac Newton (1642–1727) who first demonstrated correctly that the comet of 1680 (not Halley's) traveled in a curved orbit. Newton combined his new mathematics and his work on gravitation, which permitted exciting new insights into the movements of celestial bodies. The curved orbit that he computed for the comet of 1680 was parabolic (see figure above).

Edmund Halley (1656–1742) based much of his research on the pioneering work of his friend Isaac Newton. When Halley discovered that the comets of 1682, 1607 and 1531 made very similar appearances, he calculated matching parabolic orbits (see figure below) for all three. If the orbit was actually elliptical, it might account for even earlier apparitions and allow projections to be made for possible future returns of the same comet. With the recovery in 1758, Halley's became the first known "periodic" comet, with a seventy-six-year interval between appearances.

Further work on comets that appear only once suggested that these travel in longer paths of similar shapes where one end of the orbit brings them close to the sun (perihelion) while the other (aphelion) is far removed beyond the orbit of our farthest planet and the influence of the sun's gravity.

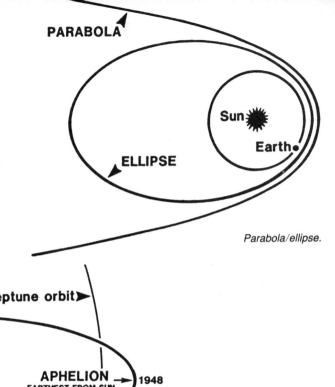

Parabola/ellipse.

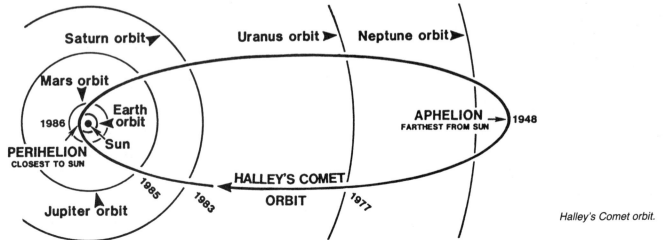

Halley's Comet orbit.

Comet West breaking apart, May 5–6, 1976. (Photo by William Liller)

In time it was learned that passing by massive planets like Saturn and Jupiter may cause disturbances in the motions of comets. They may be drawn sideways through gravitation and slowed down, or accelerated by some slingshot effect that may toss comets out into deepest space never to return.

Astronomers now believe that cometary matter may be the stuff of which our solar system was originally created some 4.6 billion years ago. They theorize that comets consist of the same most primitive components of the primordial cloud from which our sun, earth and planets are thought to be formed. Fred Whipple, the Harvard astronomer, one of today's foremost authorities on comets, concludes that comets may be "dirty snowballs," possibly with solid cores. They may even eject gases from vents in their crusty shells.

Still, no one can say with certainty what makes up a comet. We simply do not yet know. Observations have shown that comets comprise a nucleus, an enveloping coma, and tails of different and changing configurations. Far, far out in the distant absolute cold of the outermost solar system where the dormant nucleus might be observed without a surrounding coma, mere distance hides it from sight. By the time a comet nears the sun and comes within the reach of optical telescopes, the mantle of dust and gases which seem to form the coma hides it yet again and shrouds it in mystery. Where radar may signal a nucleus a mere three miles or so across, the surrounding shell of the coma can be as large in diameter as our earth, or larger. We cannot see through it.

This shroud is the mix from which comet tails are also formed. These are not atmospheric contrails like the ones streaming behind fast-flying jets or rockets. Here the radiation from the sun excites comet gases, making them glow like neon in a vacuum tube. Particles of a solar wind push them at speeds approaching a million miles an hour into tails several such hours long. Comet tails always point away from the sun. The pressure of sunlight forces the dust into a second, brighter and slower tail which bends along the path of the comet. The comet is actually leaving a trail of debris in its wake.

We have good evidence to tell us that when the earth in its own orbit round the sun crosses old cometary paths, sand-grain-sized comet remnants collide with our atmosphere, causing them to burn up in brilliant showers of meteors, a rain of shooting stars. The annual Orionid meteor showers (October 2–November 7) and the Eta (η) Aquarid shower (April 21–May 12) are believed to be related to Comet Halley.

Since even learned scientists are still theorizing not only about the makeup of comets but also about their origins, some ideas concerning their source seem particularly engaging. Here is the current theory:

An Estonian astronomer by the name of Ernst Öpik, and Jan Oort, a famous Dutch expert in the field of galaxies of stars, have put forth the concept that a shell of countless comets surrounds our sun. These original unchanged building blocks left over from the formation of our solar system dwell a thousand times beyond the orbit of Pluto, our most distant known solar planet. Occasionally a passing star may upset the gravitational balance in this huge freezer full of big and small comets, causing one or the other to start on a snail's-pace journey toward the sun. The trip may take millions of years during which they gradually accelerate, falling ever faster toward our nearest star until, held by the invisible tether of gravity, they whip around the sun at spacecraft velocities, only to be thrown back in the general direction from which they came. If in their travels they come close to massive Jupiter or Saturn, or perhaps Uranus or Neptune, these planets will tug at the visitors, altering their orbits.

Comets have been observed to break into pieces (see figure left) after a close approach to the sun, with the remaining parts of such a breakup following their own new orbits, again under the influences of the nudges, pushes and pulls of the planets and the sun.

There are three principal scenarios concerning the eventual fate of comets. The first has been actually observed, which removes it from the realm of theory: Spacecraft photographs recorded in 1979 show a member of the sun-grazing comet group as it moves ever closer to the surface of the sun, only to crash into it. The proverbial snowball in hell.

A second possible end for comets may be a collision with a planet. Numerous craters on the surfaces of planets and their satellites attest to such possibilities, not to mention the ancient so-called astroblemes on the surface of our own earth. As recently as June 30, 1908, the "Tunguska event" leveled a huge circular area of uninhabited Siberian forest in a blast heard hundreds of miles away. It was associated by contemporary observers with a bright object that had just streaked across the sky. The fact that there was no crater and that no meteoric material was later found suggests that a comet or a piece of a comet, consisting mainly of ices, may have been involved.

The third, least violent conclusion of the life of a comet might come when all its frozen liquids and gases have finally evaporated after innumerable round trips from deep space to the relative proximity of the sun. At this stage a rocky nucleus reminiscent of sterile asteroids may be left, a barren clump to drift endlessly through interplanetary space.

Many of the questions that have been posed about comets since the beginning of time may be answered during the coming 1985–86 apparition of Halley. For the first time we will be able to launch space vehicles to intercept the historic visitor, to probe the chemistry of its coma, perhaps even to photograph its mysterious nucleus close up. The Soviet Union and France have two separate missions already under way. Japan also has launched two deep-space probes. Finally, a European consortium is mounting a Halley mission named "Giotto" after the Florentine artist who painted the comet after seeing it in 1301. The United States, with its unparalleled expertise in space, is sadly conspicuous in its absence from the ranks of the pioneering nations in the quest. Paraphrasing Halley: Wherefore when the comet returns according to our predictions about the year 2061, impartial posterity will not fail to remember that in 1986, when technologically advanced nations were first able to soar out into space to greet the historic comet, America was not among them.

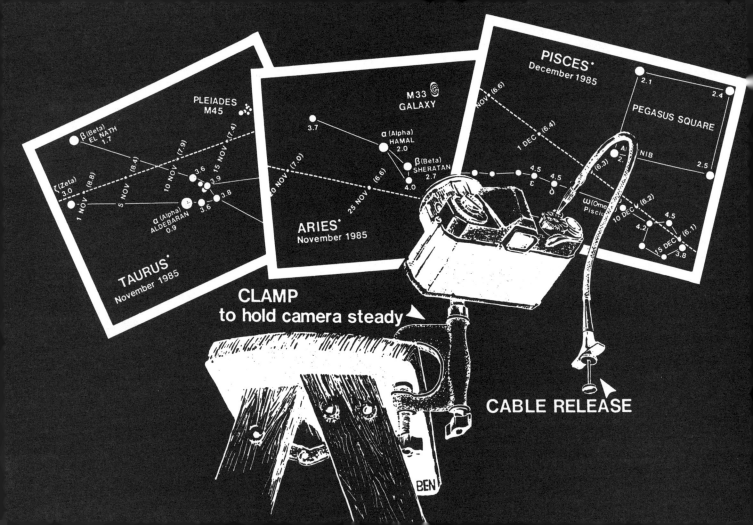

4. Best Ways to Find the Visitor Early

To catch sight of any comet for the first time is an exciting event with its own reward. The early rediscovery of Halley's famous treasure will combine challenge with a special kind of satisfaction. It will be exhilarating to be among the very first to see this ancient solar migrant.

Early recovery of Halley's Comet in November 1985 will be possible for anyone with enough determination, but will be challenging and possibly elusive. December 1985 should definitely bring it within your reach. No, you do not have to freeze to death standing in the cold night trying to find the distant nucleus. You will be using a foolproof and easy method which has been employed by professional astronomers since the beginning of this century. It is so simple and elegant that you will wonder why you did not think of it yourself.

The concept is based on the known fact that a comet almost always moves, while the stars among which it travels stay put in their places. The system you will adopt requires only a simple camera. To find Halley's Comet easily and quickly once it comes into sight, all you need to do is to go out for a few minutes to shoot a night picture of the *area* where we know—even now—the comet will be on a given date. Then, a day or two, even three nights later (weather permitting), you shoot a second picture of the same *area*. If you follow the recommendations in this book (see photography, Chapter 11) your film will now show two almost identical photos of a sky region with stars. The two photographs will also have recorded the comet, but its position will have changed in relation to the background stars.

Next you compare the photos to find which images are fixed stars and which one is the moving comet. All this can be done leisurely in the comfort of a warm room. Comparing the pictures star for star would be tedious. A simple comparator allows both frames to be viewed through a pair of magnifiers at the same time by two ordinary night lights. Once the stars in the two photos have been aligned stereoscopically to appear as single images, the comet (the only object that will have changed its position) will reveal itself instantly by jumping back and forth when you switch one of the pair of night lights rapidly on and off and on again, etc. You will be using a "STEBLICOM" discovery box which professional astronomers call a "blink comparator" (see Chapter 13).

Detailed photographs of the constellations Taurus, Aries and Pisces, which follow, show where the comet will journey between November 1 and December 15, 1985. All three areas have bright target stars in them which are visible even from cities. It is at these stars that you will be aiming your camera to photograph the *area* on any of the dates shown. Also in these three constellation maps will be an indicator star showing what magnitude the comet will have while in the particular starfield. (See pages 30, 34 and 38.)

You should practice your skills in simple astrophotography in advance of the comet's arrival. On any clear night you can try to record the faintest stars possible. Only a few clues are needed. After that it is a matter of how long to leave the shutter open. (See Chapter 11.)

There is a wonderful bonus to the method we will be using to find and track the comet. The human eye can see only a faint star for the instant (about $\frac{1}{10}$ second) during which an area is actually being scanned. A distant comet, unless it is bright and in easy sight, is therefore hard to find. Not so with a camera. The highly sensitive films available today allow the camera to collect starlight over a period of time. Film keeps on accumulating starlight or the merest glow of a comet during all the time that the lens is held open. You can actually gather 15 seconds' worth of faint cometlight. In the resulting photographs you will see hundreds of times more than the naked eye ever could see.

As soon as the comet starts growing a tail this advantage becomes doubly

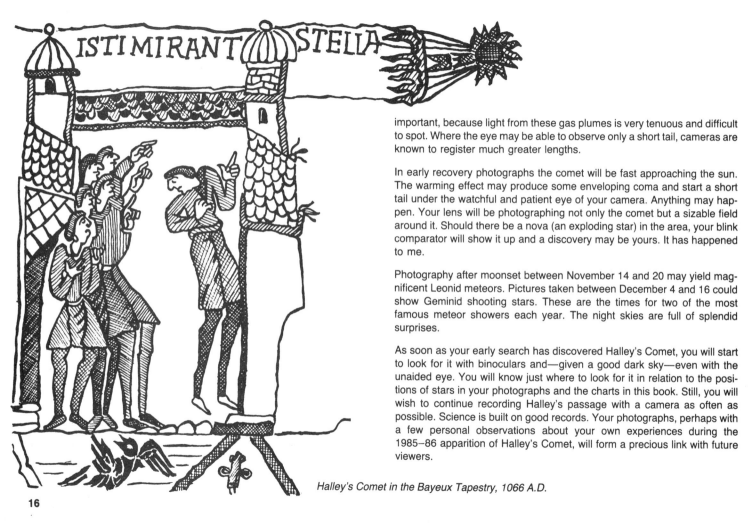

important, because light from these gas plumes is very tenuous and difficult to spot. Where the eye may be able to observe only a short tail, cameras are known to register much greater lengths.

In early recovery photographs the comet will be fast approaching the sun. The warming effect may produce some enveloping coma and start a short tail under the watchful and patient eye of your camera. Anything may happen. Your lens will be photographing not only the comet but a sizable field around it. Should there be a nova (an exploding star) in the area, your blink comparator will show it up and a discovery may be yours. It has happened to me.

Photography after moonset between November 14 and 20 may yield magnificent Leonid meteors. Pictures taken between December 4 and 16 could show Geminid shooting stars. These are the times for two of the most famous meteor showers each year. The night skies are full of splendid surprises.

As soon as your early search has discovered Halley's Comet, you will start to look for it with binoculars and—given a good dark sky—even with the unaided eye. You will know just where to look for it in relation to the positions of stars in your photographs and the charts in this book. Still, you will wish to continue recording Halley's passage with a camera as often as possible. Science is built on good records. Your photographs, perhaps with a few personal observations about your own experiences during the 1985–86 apparition of Halley's Comet, will form a precious link with future viewers.

Halley's Comet in the Bayeux Tapestry, 1066 A.D.

5. How to Find the Host Constellations

Knowing that Halley's Comet will travel through Taurus and Aries into Pisces and then Aquarius is important and useful advance information for which we are ultimately indebted to Edmund Halley. For the exact night-to-night positions during the 1985–86 passage we owe much to brilliant scientists like Dr. Donald Yeomans, who, with the computers at the Jet Propulsion Laboratory in Pasadena, California, forecast accurate positions several years in advance. His refined calculations suggested to astronomers worldwide where and when to start looking for the comet. The initial "recovery" photograph in October 1982 (see page 6) owed much to his work. (Recovery means the rediscovery of an object seen years or decades earlier.)

But what good is it for us to know that we get early photos of the comet in Aries and Pisces in November–December 1985 and our first visual sighting in January 1986 "in Aquarius" when we do not know where to find the constellations that will play host to our historic visitor? Before we proceed from early recovery to first encounters (trying to observe the comet visually with binoculars or the unaided eye), let us find the host constellations.

Blinking night lights will help us to recover the comet. We will also employ completely new methods to find the constellations when we bend ordinary wire coathangers into rectangular Starframes™ to isolate specific areas in the sky for search and study. For complete information on using Starframes™ to find constellations, see *Starwatch,* by Ben Mayer (Perigee Books, 1984). Chapter 6 of this book gives a short discussion. By holding our starframes against the night skies some 15–20 inches from our eyes, we will actually outline the starfields that are pictured in this book.

An important additional benefit that comes from using starframes is that by aiming them at selected bright target stars and holding these stars in the center of the field of view, you can readily visualize the area a standard camera photographs. This is very important in the night sky, where there are none of the reference points that guide us during daytime photography.

Finally—and this is perhaps the most important use for starframes—they will help us visualize the angles at which we must hold them and aim our cameras in order to make the constellation photographs correspond to the actual positions of the stargroupings in the sky. As can be seen on pages 8 and 9, the constellations in which Comet Halley travels follow quite literally a curved path which pretty well corresponds to the journey that the sun also travels, especially in winter, when it never seems to get very high in the sky. This path, which is called the "ecliptic," begins near the southeast point on the horizon, rises to the twelve-o'clock midday (or midnight) position and ends near the southwest point, again on the horizon. The simple fact is that while rounding the sun Halley's Comet will logically be found in six of the twelve constellations of the Zodiac where the sun also seems to travel.

When we say that the sun *seems* to travel daily from east to west in its journey around Planet Earth, we recognize what Copernicus and Galileo taught us four hundred years ago: The sun really stays in one place. It is the earth which turns and moves around the sun. This creates the *illusion* of solar motion. It is this very same turning of our earth which places us under different constellations in the carousel of the Zodiac at different times of the night and the year.

To simplify astronomical observations and our search for Comet Halley it is very important that we be always aligned with the axis on which our earth turns. This axis extends from the North Pole through the very center of our globe to the South Pole. If we lie down with our head to the north, feet pointing south, we will almost be aligned with this axis. To standardize our Halley watch it will be more convenient to stand (or sit) in a "polar-aligned position": back to the north, facing south, because *all of the comet's appearances will be near the southwestern or southeastern or above the southern horizons.* When in doubt about directions, use a small compass to orient yourself (see page 18).

The primary purpose of these pages is to help us discover, observe, even

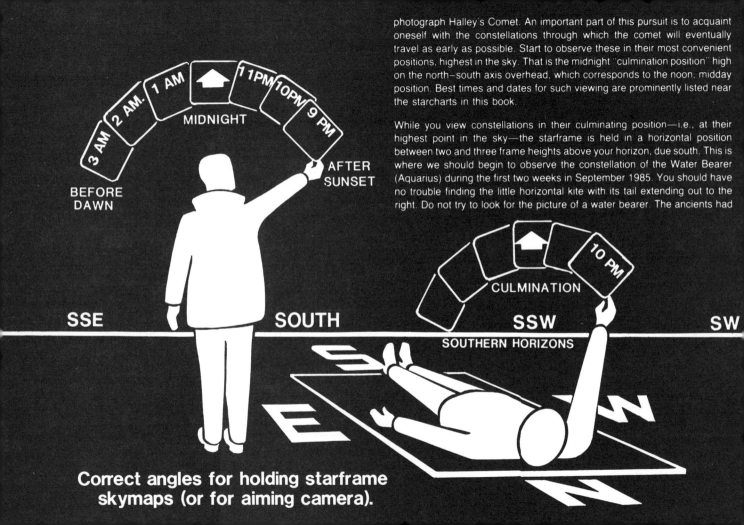

photograph Halley's Comet. An important part of this pursuit is to acquaint oneself with the constellations through which the comet will eventually travel as early as possible. Start to observe these in their most convenient positions, highest in the sky. That is the midnight "culmination position" high on the north–south axis overhead, which corresponds to the noon, midday position. Best times and dates for such viewing are prominently listed near the starcharts in this book.

While you view constellations in their culminating position—i.e., at their highest point in the sky—the starframe is held in a horizontal position between two and three frame heights above your horizon, due south. This is where we should begin to observe the constellation of the Water Bearer (Aquarius) during the first two weeks in September 1985. You should have no trouble finding the little horizontal kite with its tail extending out to the right. Do not try to look for the picture of a water bearer. The ancients had

3 AM
2 AM.
1 AM
MIDNIGHT
11PM
10PM
9 PM
BEFORE DAWN
AFTER SUNSET

SSE
SOUTH

CULMINATION
10 PM
SSW
SW
SOUTHERN HORIZONS

Correct angles for holding starframe skymaps (or for aiming camera).

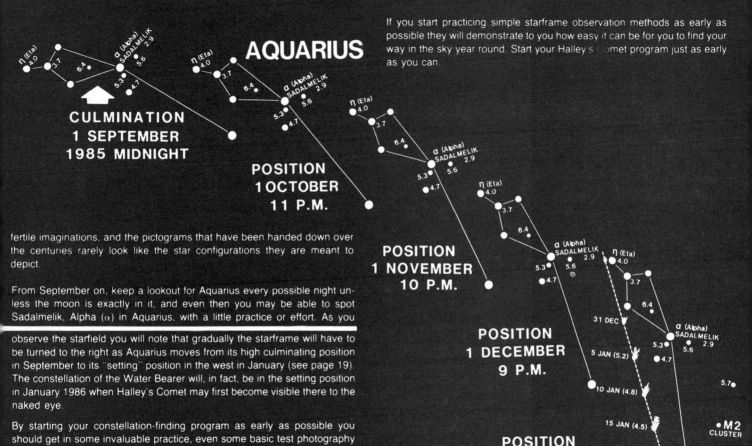

If you start practicing simple starframe observation methods as early as possible they will demonstrate to you how easy it can be for you to find your way in the sky year round. Start your Halley's Comet program just as early as you can.

AQUARIUS

CULMINATION 1 SEPTEMBER 1985 MIDNIGHT

POSITION 1 OCTOBER 11 P.M.

POSITION 1 NOVEMBER 10 P.M.

POSITION 1 DECEMBER 9 P.M.

POSITION 1 JANUARY 1986 8 P.M.

fertile imaginations, and the pictograms that have been handed down over the centuries rarely look like the star configurations they are meant to depict.

From September on, keep a lookout for Aquarius every possible night unless the moon is exactly in it, and even then you may be able to spot Sadalmelik, Alpha (α) in Aquarius, with a little practice or effort. As you observe the starfield you will note that gradually the starframe will have to be turned to the right as Aquarius moves from its high culminating position in September to its "setting" position in the west in January (see page 19). The constellation of the Water Bearer will, in fact, be in the setting position in January 1986 when Halley's Comet may first become visible there to the naked eye.

By starting your constellation-finding program as early as possible you should get in some invaluable practice, even some basic test photography under warmer summer skies. Then you can keep your late-fall or winter observing and shooting times as short as possible: just a few quick minutes outdoors before you continue to conduct your actual search indoors in warm comfort.

6. Equipment and Methods

In order for us to find Halley's Comet the old-fashioned visual way, very little equipment is required and what we need to start is neither expensive nor complex. The most important tool may well be an ordinary compass that will help us find north. This, in turn, will enable us to align ourselves with the north–south axis of the earth as explained in the previous chapter.

Next we need a wire coathanger to bend into a starframe some ten inches wide by eight inches high. This will allow us to frame our evening, morning or night sky areas for an orderly search.

We should also have a flashlight in which the small bulb has been painted red with nail polish, or covered with a tiny piece of red cellophane. (Candy wrappers will do just fine.) This is to read our starcharts and to take notes by dim and safe light.

Even though it is a real advantage to have a telescope, a camera is a far more useful comet-finding tool. This will be further explained in Chapters 11 and 13.

The most complex and precious optical instruments needed for our comet search—or, for that matter, for any astronomical observing program—are our eyes. Such priceless and sensitive viewing equipment deserves the most careful preparation and conditioning. Before we attempt any nighttime observations it is absolutely necessary that we adapt our eyes to the dark. This important step is mostly overlooked or completely forgotten.

Anyone who has ever walked from a bright cinema lobby into a darkened theater knows that at first one may not even be able to find a vacant seat. Experience has taught us that after a few minutes our eyes adjust to the conditions of darkness and we are gradually able to see what was hidden to us before.

How to make a Starframe.

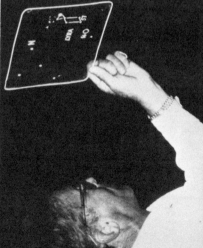

Use an inexpensive compass to orient yourself.

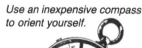

How to find constellations with a Starwatch Starframe.

Once our eyes have reached this state of high sensitivity, which also involves subtle chemical changes within the eye, even one brief look at a bright light or a television screen can spoil it, requiring a new period of readjustment. The only type of light that will not offend our vision or dull our sensitivity is dim red light. That is why astronomical observatories are always bathed in a gentle red glow.

You must shield your flashlight with red cellophane as described above or with a piece of red or crimson cloth. Red plastic bottle caps can also be taped over the bulb or the lens to dim the white glare. This, then, will be the shaded light source by which you read your compass and your skymaps. You should take notes by red light and even work your camera by its glow.

Always select an observing site away from lights, ideally with as wide a view of southern horizons as possible. The farther you can move away from city lights, the darker your skies and the better your views will be. Take loved ones or invite children along for memorable nights of starwatching and comet searching.

In astronomy this wonderful natural eye adjustment is of the greatest importance. Time and again various conditions are blamed for one's inability to see stars or a faint comet. Impatience is almost always the cause. It is absolutely impossible to go from a bright room into the night and to make observations immediately; in fact, it is a physical impossibility. Human eyes need at least ten or even fifteen minutes to become accustomed to the dark. Experienced professional astronomers allow a full half hour to adapt to darkness. City dwellers, most of all, should rest their eyes awhile under the night skies. They will then be able to see stars of much fainter magnitude than they considered possible.

What actually happens during the time of conditioning is that the pupils in our eyes gradually dilate so that when they are at their largest they admit the maximum amount of light through the iris for best seeing. This is like opening the "f" stop on the lens of a camera to its widest open setting. We want to achieve maximum aperture for our eyes for best night viewing, just as we will open our camera lenses as far as possible for night photography.

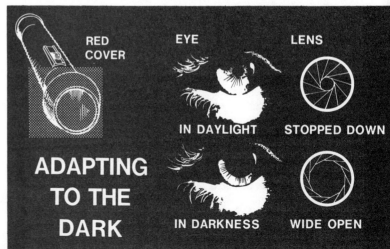

7. Star Atlases, Skymaps, and Guides

Just as we have atlases and maps to help us find our way on the surface of our earth, there are star atlases and skymaps to guide us around the heavens. In this chapter we show precise, detailed but amazingly easy-to-understand maps of the constellations through which Comet Halley will travel between November 1985 and May 1986. These handy starmaps will serve us well when we study them carefully.

Another guide, *Starwatch* (Ben Mayer, Perigee Books, 1984), will be of particular help to you. It shows all twelve Zodiac constellations together with another dozen of the most popular year-round starfields. The scale of *Starwatch* is large and it was specially designed to present a completely new method for finding constellations. You can trace principal stars from the book right onto your starframes. When you line these traced stars up with their counterparts in the sky at the culmination times given in the book, you can find your way around the heavens quite easily. There is also additional information in *Starwatch* for those who want to further explore the night sky and learn more about astrophotography.

The following pairs of guide and atlas pages have been designed to help Northern Hemisphere observers to find Comet Halley between November 1985 and May 1986 and to shed more light and color on the special visitor. They help you follow the comet through six famous constellations of the Zodiac and out again through less prominent starfields.

The names of the constellations are boldly listed first, followed by an explanation of the comet's passage through the area. The brightest stars shown in the specific starmaps are listed. Most of these are visible even from dark areas near light-polluted cities once your eyes have properly adapted to the dark (see page 21). All of these principal stars can be readily photographed with 15–20-second tripod exposures (see pages 63–67). Star names and their translations from the Arabic (AR) or the Greek (GK) are given. By using the Greek letters shown in combination with the name of the constellation you can identify each star scientifically. Thus, Aldebaran, the brightest star in Taurus, can also be referred to as "Alpha (α) in Taurus"; and Sheratan, the second brightest star in Aries, would be "Beta (β) in Aries." (See the Greek alphabet below.)

THE GREEK ALPHABET

α	ALPHA	ν	NU
β	BETA	ξ	XI
γ	GAMMA	ο	OMICRON
δ	DELTA	π	PI
ε	EPSILON	ρ	RHO
ζ	ZETA	σ	SIGMA
η	ETA	τ	TAU
θ	THETA	υ	UPSILON
ι	IOTA	φ	PHI
κ	KAPPA	χ	CHI
λ	LAMBDA	ψ	PSI
μ	MU	ω	OMEGA

Magnitudes (brightnesses) of the main stars are given, also their distances from earth in light-years (1 LY = 5,865,696,000,000 miles—that is, nearly 6 trillion miles). Concerning the changing visual magnitudes of the comet, it will be difficult to compare them with stars because comets usually have a fuzzy appearance. In photographs such comparisons are easier to make.

Much is being written about the eventual magnitude of Comet Halley. The first projections indicated rather dim appearances. Then estimates were revised in late 1983 and early 1984 to suggest a brighter comet than origi-

About a dozen comets of varying brightness become visible each year.

Comet Kohoutek, January 1974. (Photo by Dennis di Cicco)

Comet West, March 1976

nally anticipated. For our first glimpse of the comet this is of no concern whatsoever. Rather, it is a matter of how faint a star we can photograph so that we can trap the comet on our films at the earliest possible moment for recovery. There is absolutely nothing we can do about making the comet brighter. Endless resources, however, are at our disposal to capture faint cometlight as soon as it becomes available, to collect its glow and enhance it (see Chapter 12).

The scale of magnitudes of stars becomes clear when we view the charts of Taurus. Aldebaran, Alpha (α) in Taurus, has a magnitude of 0.9; El Nath, Beta (β) in Taurus, checks in at 1.7, and Alcyone, Eta (η) in Taurus, is magnitude 2.9. For simplicity's sake let us call these 1st-, 2nd- and 3rd-magnitude stars. Please note that *the lower the number, the brighter the star.* Under good and dark skies the unaided eye can see stars to 6th magnitude. Our cameras can do much better yet.

They can capture objects to the 10th magnitude. The left-hand negative prints show stars to 9th magnitude and beyond. An approximate indication is given of the anticipated brightness of the comet in relation to other stars in the starmaps of Taurus ($\approx$ mag. 8), of Aries ($\approx$ mag. 7), and of Pisces ($\approx$ mag. 6).

One final word before we delve into our skymaps. Do not hesitate to seek help in your comet quest. Call your nearest planetarium or a college with an astronomy or earth-sciences department to find out where and when the local amateur astronomers meet. Attend a meeting and take this book with you. You will find starlovers in all walks of life, and they will be proud and happy to answer your questions. You may even be invited to a "star party" on some weekend when there is no moon and the skies are dark. Go and join the fun. There you will meet kindred souls who will eagerly show you their telescopes and let you look through them for a first look at the magic of the night sky. You will be shown how to relate the starcharts in these pages to the stars in the sky. Someone may give you more firsthand suggestions on how to start out in photography. You may even want to become a member of an astronomy club. *No experience is required.* Membership fees

M45, Pleiades Cluster in Taurus. *M8, Lagoon Nebula in Sagittarius.* *M20, Trifid Nebula in Sagittarius.*

are always low. All you need to bring is a love for the stars and curiosity. *You do not need to own a telescope to start.* In fact, you may learn there how to build your own. Once you have looked through a 'scope you may become hooked on the stars.

Soon you will want to subscribe to one of the amateur publications (see "Resources," page 77). Many of them will discuss Halley's Comet and will print photographs. Such monthly publications will also feature the constellations that are conveniently placed for observation during any given month, and you can key the starcharts in this book to those maps.

The following starmaps detail the exact path of Halley's Comet. The maps are always presented in pairs. The left photographs show faint detail of long exposures, the right demonstrate what beginners can shoot. All of these photographic maps were recorded with a standard 35mm. SLR (single-lens

reflex) camera while constellations were at their highest under a black country sky. Exact times and dates for such "culminations" (when the starfields are placed for most convenient recognition) are listed near the right-hand photographs. The "fastest" color films were used (ISO 1000 or faster) and the camera was mounted on a tripod during 15–20-second exposures for the photos on the right pages.

The left-hand pictures are so-called "negative prints" which were made by using color slides as negatives for black-and-white printing. (Projection of color slides throws planetarium-sized images on any large white wall.) Again fastest films were used (ISO 1600 and up). For such 4–8-minute exposures a standard 35mm. SLR camera was mounted on a polar-aligned motor drive, as shown on page 69.

In all cases when the films were sent out for processing, "push development" was requested. This makes all films more light sensitive even *after* photographs have been taken. It permits the recording of faintest stars and the earliest finding of a comet.

Chapter 11, "Collecting Cometlight," will discuss simple astrophotography in greater detail and will tell you how to shoot starmaps like the ones in this book for yourself. Photography is not limited to dark-sky locations, although best results will be obtained under darkest possible skies. Start early to test for yourself the optimum results you may obtain from your own most convenient locations. It is for this purpose that space has been provided for keeping accurate records for photography in each starfield. Add pages as needed. Your notes will become your most invaluable aid, a priceless starlog for the future.

To help relate our skymaps to the celestial sphere and to other starcharts, a vertical line and a horizontal one are superimposed on each of the negative prints. These indicate the "nearest major intersection" in the gridwork used to mark positions in the sky. The vertical lines mark the hour circles that divide the heavens into twenty-four imaginary melon-slice shapes. These correspond to the lines of longitude on the surface of our earth. The horizontal lines mark the angles of declination of the specific starfields. When the areas are above the celestial equator the degrees are noted with a plus (+) sign. When they are below this imaginary sky equator (which extends skyward from the equator of our earth), the degrees are preceded by a minus (−) sign. The horizontal hoops of declination are similar to the lines of latitude on our earth. Together, vertical and horizontal lines are called coordinates. By detailing the hour, the minute and the second as measured horizontally and the angle of declination in vertical degrees, you can pinpoint exactly every star, each changing position of Halley's Comet and all the nebulae, clusters and galaxies in the heavens.

Also shown in the "negative prints" are several "M" objects. There are 110 of these "Messier objects" in the northern skies. They were named after a famous French astronomer who was possessed by a passion for comets. Only about fifty comets were known when Charles Messier (1730–1817)

began looking for more from above the roofs of Paris, recording his observations by lantern light. In 1754 Messier became a clerk at the marine observatory in Paris, which was housed in the Hôtel de Cluny, a residence. Messier became so intensely involved with the search for comets that King Louis XVI called him the "comet ferret." Vanity may have been the reason the young astronomer spent so much time scanning the sky for the elusive newcomers, because by this time it had become the custom to name comets after their discoverers. Over his lifetime Messier claimed to have found some twenty-one comets of his own. Modern astronomy has revised this figure downward and credits him with possibly fifteen original discoveries.

M33, Spiral Galaxy in Triangulum.

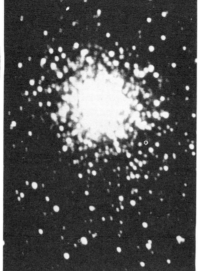

M2, Globular Cluster in Aquarius.

M22, Globular Cluster in Sagittarius.

M55, Globular Cluster in Sagittarius.

M68, Globular Cluster in Hydra.

The so-called "Messier objects" were a byproduct of his intensive searches. Galaxies, nebulae and star clusters were at first mistaken by Messier for comets when he glimpsed such hazy patches through his primitive eighteenth-century telescope. When he noticed that the objects did not move from night to night—as comets should—the Frenchman started a list giving the exact positions for such faint targets, to spare himself and future comet seekers the trouble of being misled by blurry spots in the night sky.

Today even amateur telescopes will resolve what Messier could see only faintly. M8 is revealed as the magnificent Lagoon nebula, birthplace of new stars (see page 24). The nearby Messier object M22 is a glorious globular star cluster that contains over seventy thousand stars, many larger than our own sun (see above).

It was during his unending quest for comets that Charles Messier spotted the comet which Edmund Halley had predicted would return every seventy-six years. As is so often the case, Messier, the professional astronomer, made his discovery almost a month after Johann Georg Palitzsch, a German amateur, had sighted the comet on Christmas night in 1758.

The lure of the sky and the passion for discovery that kept observers riveted to their telescopic eyepieces even during long winter holiday nights today involves the modern camera, which allows for searches to be conducted with greater comfort and ease. But the truth is that we still know very little about comets. We do not know where they come from or what makes any of them arrive so unexpectedly. We do not know what they look like close up. The mystery stays alive. Edmund Halley merely unlocked the secret of their paths, not the path to their secrets.

This book shows all the Halley host constellations—between November 1985 and May 1986. To familiarize yourself better with the Halley host constellations, study the star images in the negative prints and their Greek-letter names in relation to the pictograms in the right-hand photographs. Then connect the constellation stars with colored lines to create your own pictograms. If you use a red pencil to connect the stars, the red lines will vanish under a red flashlight at night. The red light source will filter out the red lines and make them seem to disappear. That way your maps will be unmarked for easiest night viewing.

Usually in a book on astronomy or in a sky atlas, constellations are presented in the order in which they culminate. They are offered here in the order in which Halley's Comet may be seen in them: Taurus, Aries, Pisces, Aquarius, Capricorn, Sagittarius and Hydra. This completely reverses the customary order. It also means that as we prepare for the visitor by familiarizing ourselves with the *host constellations* in advance, we have to use the following section backward. This need not present any problems, because the dates and times of culmination are always prominently shown.

For observing convenience you may want to start exploring the constellations in their order of culmination. You may wish to observe Sagittarius the Archer in August 1985. You can spend the summer months roaming through this region, perhaps the most beautiful of the starfields, continuing through Capricorn into Aquarius in autumn 1985. It will be in Aquarius the Water Bearer that we should make first naked-eye contact with Comet Halley. You would therefore want to start watching the kite-shaped pictogram in Aquarius in September 1985 and follow it westward and to its setting position in January 1986. Pisces, Aries and Taurus rise in sequence in the east during the summer of 1985, to culminate in the same order in September, October and November. By that time you should already have taken your first constellation photographs for comparison on the off chance that you might have caught Comet Halley earlier than you had expected. Perhaps you had a chance to improve your photographic results through the use of a 135mm. telephoto lens or the acquisition of a drive system on which to mount your camera.

Space is provided for you to keep notes concerning your observations. Any records in connection with photographs that you may take are particularly valuable when you receive your pictures back from processing. You simply cannot remember exposure times or camera settings unless they are *written down*. Since you may shoot certain regions for practice long before photographing them again for your tries at early recovery, it is important that you write down all the data not only in this log but on each slide or the back of each print to establish your own best results. Then, when weather conditions are the same, and you know that the comet is actually in the field you are photographing, you only have to repeat your optimum settings. This book will become a valuable record and its findings your best teacher. Read Chapter 11, aim your camera and shoot. You have nothing to lose and a view of Halley's Comet to gain.

Charles Messier, 1730–1817.

TAURUS

To get a good start on this constellation, you may want to get up early during October 1985 and have your first look at this splendid starfield. If you are using a starframe, prepare it the night before and have everything ready so that you can get dressed in the dark. Don't turn on any bright lights, because after sleep your eyes will be perfectly dark-adapted.

There is no point trying any photography yet, because at about the time that Taurus will be at its highest in the sky there will be the first beginnings of the dawning of the new day. This is called the "beginning of astronomical twilight" after which the sun will be less than eighteen degrees below the eastern horizon. Still, you will get a fine look at the Pleiades, also known as the "Seven Sisters." You should see about six or seven stars there. With binoculars, ten times as many may be spotted. This grouping is an "open" cluster lying in the Frisbee-shaped disc of our galaxy. It is also known as a "galactic cluster."

Aldebaran, Alpha (α), is the brightest star in Taurus, and you can align your starframe with it and with the Pleiades. El Nath, Beta (β) in Taurus, is off the chart at the top left. It forms the tip of the northern horn and can be easily found. The Messier object M1 is a very faint telescopic remnant of a star explosion that occurred over nine hundred years ago. All that remains is a cloud of glowing gas which, even today, expands at the rate of some 700 miles per second.

Where on the first of October in 1985 you will be able to watch the almost full moon set in the west just as the sun rises in the east, conditions by midmonth will have improved to new moon, when the lunar disc will rise and set with the sun, leaving the sky dark all night. This will afford fine opportunities for photography of the constellation, even though the comet will not make its entrance into Taurus until November. Then you will shoot Taurus again to try to capture the comet. Do not be discouraged if you fail to find or capture it just yet. You will gain invaluable experience with each shooting session.

THE Bull — TAU / Tauri

L.Y.	BRIGHTEST STARS		MAG.
68	α ALPHA	ALDEBARAN — PLEIADES' FOLLOWER (AR)	0.9
130	β BETA	EL NATH — PUSHER, WITH HORNS (AR)	1.7
130	ε EPSILON	AIN — EYE OF THE BULL (AR)	3.6
490	ζ ZETA	ZETA TAURI (GREEK LETTER + POSSESSIVE)	3.0
240	η ETA	ALCYONE (BRIGHTEST STAR IN PLEIADES)	2.9

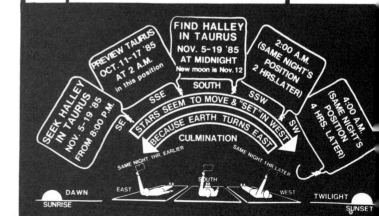

PHOTOGRAPHIC RECORD | TAU | Date | VISUAL RECORD

Film:		Type:		ISO	
				ISO	

CAMERA:			CLAMP/TRIPOD EXPOSURES, SECS.	PIGGYBACK/DRIVE EXPOSURES, MINS.	
LENS:		TAKEN			

DATE:		SHOT#	AT HRS.	USE APPLICABLE COLUMN								
	FILM#			10				1M				
					15				2M			
						20				4M		
							40				8M	

DATE:		SHOT#	HRS.									
	FILM#											

DATE:		SHOT#	HRS.									
	FILM#											

DATE:		SHOT#	HRS.									
	FILM#											

RESULTS AND NOTES

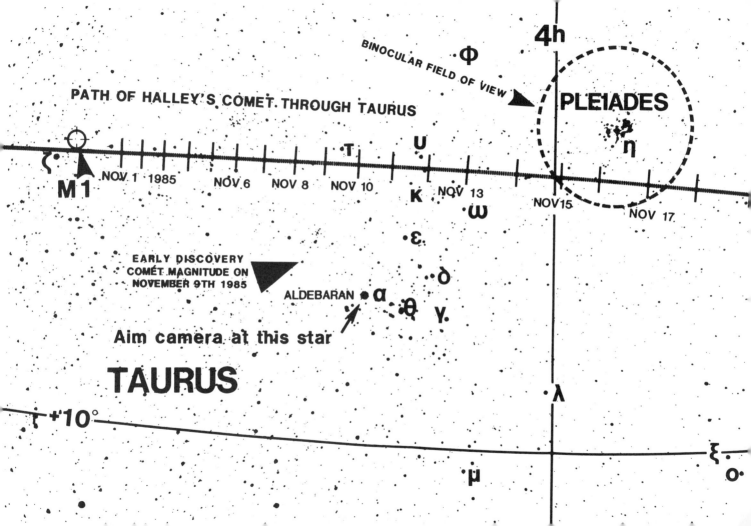

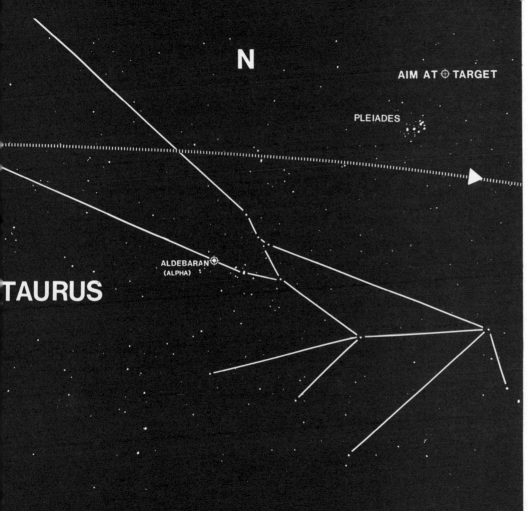

N

AIM AT ⊕ TARGET

PLEIADES

ALDEBARAN ⊕
(ALPHA)

TAURUS

Culmination dates and times: when the constellation will be at its highest, some 20° to 40° above the southern horizon.

TAURUS

15 SEP.	5 A.M.
1 OCT.	4 A.M.
15 OCT.	3 A.M.
1 NOV.	2 A.M.
15 NOV.	1 A.M.
1 DEC.	MIDNIGHT
15 DEC.	11 P.M.
1 JAN.	10 P.M.
15 JAN.	9 P.M.
1 FEB.	8 P.M.
15 FEB.	7 P.M.
1 MAR.	6 P.M.

For Daylight Saving Time, add one hour to the times shown above during the months from April through October.

ARIES

Both Aries and Taurus (adjoining constellations, with Aries on the right) are well placed for observation and photography an hour or two apart near midnight during the week of November 12 to 19, 1985. At the beginning of this period there will be no moon (the moon will be new, rising and setting with the sun, and will be lost during the day in the solar glare). About November 19 the first-quarter moon will have set before midnight, leaving a moonless sky for us to start to take pairs of photographs, to try our hand at blinking (see Chapter 13). Do not be disappointed, though, if you cannot yet find Halley's Comet. It is still very faint. Unless your practice photography earlier in 1985 has established that you can reach stars of magnitude 8, you will have to wait until December before making your own recovery of the comet.

Be on the lookout in your photographs of the Aries region for the triangle just north of Hamal, Alpha (α) in Aries. The trio of stars make up the relatively unknown constellation of Triangulum. This area contains a noted galaxy which, like the Andromeda galaxy and our own Milky Way, contains thousands of millions of stars. It is an "island universe" some 2.3 million light-years away from us. Charles Messier gave it the number 33. Although the object is diffuse, large and fuzzy, it has a total light output equal to a star of about 5th to 6th magnitude. It could register on some of your photographs.

If in order to take your photographs you drive out to a dark region where the sky is filled with a myriad of stars, a binocular look at the position of M33 may reveal this cosmically close neighbor. Astronomers regard the "spiral in Triangulum" as a member of our "local group" of galaxies.

Test photography of the constellation of Aries before December 1985 will acquaint you with the field and the faintest stars that your equipment and best location allow you to record. If you can capture a star as faint as the "comet magnitude" star marked on page 34, Halley's Comet will be in your November–December photographs.

RAM + TRIANGLE — ARI
Arietis

L.Y.	BRIGHTEST STARS	MAG.
85	α ALPHA **HAMAL** — THE RAM (AR)	2.0
46	β BETA **SHERATAN** — THE TWO SIGNS (AR)	2.6
150	γ GAMMA **MESARTHIM** — SERVED ONES (HEBREW)	5.0
170	δ DELTA **BOTEIN** — LITTLE BELLY (AR)	4.5

SAME NIGHT'S POSITION 2 HRS. PRE- (BEFORE) CULMINATION. Note hourly motion of the earth

PREVIEW ARIES SEPT. 11–17 '85 AT 2 A.M. OR OCT. 11–17 '85 AT MIDNIGHT

FIND HALLEY IN ARIES NOV. 19–30 '85 AT MIDNIGHT. Full moon a problem near Nov. 27

SAME NIGHT'S POSITION 4 HRS. POST- (AFTER) CULMINATION

SE — SSE — SOUTH — SSW — SW

STARS SEEM TO MOVE & "SET" IN WEST BECAUSE EARTH TURNS EAST

CULMINATION

ARIES

DAWN — SUNRISE — EAST — SOUTH — WEST — TWILIGHT — SUNSET

PHOTOGRAPHIC RECORD — ARI | Date | VISUAL RECORD

| Film: | | Type: | | ISO | |
| | | | | ISO | |

CAMERA:

| LENS: | | TAKEN | CLAMP/TRIPOD EXPOSURES, SECS. | PIGGYBACK/DRIVE EXPOSURES, MINS. |

	SHOT#	AT HRS.	USE APPLICABLE COLUMN								
DATE: / FILM#			10				1M				
				15				2M			
					20				4M		
						40				8M	

	SHOT#	HRS.									
DATE: / FILM#											

	SHOT#	HRS.									
DATE: / FILM#											

	SHOT#	HRS.									
DATE: / FILM#											

RESULTS AND NOTES

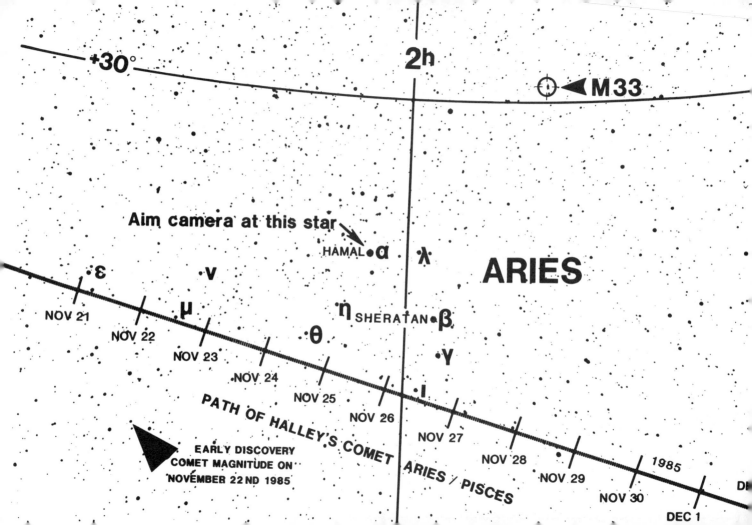

N

AIM CAMERA AT HAMAL

HAMAL
(ALPHA)

SHERATAN
(BETA)

ARIES

Culmination dates and times: when the constellation will be at its highest, some 20° to 40° above the southern horizon.

ARIES

15 AUG.	5 A.M.	
1 SEP.	4 A.M.	
15 SEP.	3 A.M.	
1 OCT.	2 A.M.	
15 OCT.	1 A.M.	
1 NOV.	MIDNIGHT	
15 NOV.	11 P.M.	
1 DEC.	10 P.M.	
15 DEC.	9 P.M.	
1 JAN.	8 P.M.	
15 JAN.	7 P.M.	
1 FEB.	6 P.M.	

For Daylight Saving Time, add one hour to the times shown above during the months from April through October.

PISCES

The constellation Pisces was called "Gemini Pisces" by the Romans, because there are two fishes. Our search for Comet Halley will be concentrating on an area best found by locating a bright star belonging to adjoining Pegasus. The square of Pegasus with its easy-to-find four stars is cradled by the fainter stars of the constellation Pisces. To find the bright star Algenib, Gamma (γ) in Pegasus, place the square of Pegasus in the top right corner of your starframe. Algenib will then be in the middle of the area through which Comet Halley travels. The comet's path lies just south of this conspicuous rectangle.

Pisces offers you an opportunity for finding and photographing the constellation well in advance of the passage of the comet through the Pisces starfield. You can start in October and keep the area in sight through November, so that by the time the good moonless nights roll around you should be ready with best exposure times and correct targets at which to aim your camera. Try any evening between December 5 (when the moon is in its last quarter and rises after midnight) and December 12 (new moon). In the nights following you can still trap the comet and watch the faintest first traces of the waxing crescent showing up in some of your photographs. Just keep aiming at the Pegasus star Algenib, Gamma (γ), or possibly just a tiny bit below it to be sure the entire lower fish will be in your photo. All through these times the comet will be at about magnitude 6 and brightening. I have taken test photos where 6th-magnitude stars show up very well in 30-second exposures taken with a standard camera on a tripod. Even if the comet has a fuzzy appearance, by now it should register as a distinct light source. It may reveal itself as having a bluish color which will further set it apart from the surrounding star-dots.

From now on it is open season on the comet. Moon or no moon, you should want to try to keep shooting and blinking so that when you have recovered the historic traveler you can begin to seek it out with binoculars and hopefully soon thereafter with the naked eye. Be sure you map its progress with marks and notes in your comet log. Please keep your records accurately. You—and others in the future—will treasure them.

THE Fishes | PSC
Piscium

L.Y.	BRIGHTEST STARS	MAG.
130	α ALPHA **ALRISCHA** THE ROPE–KNOT (AR)	3.9
125	γ **GAMMA PSC**	3.8
450	η **ETA PSC**	3.7
150	ω **OMEGA PSC**	4.0
180	τ **TAU PISCIUM**	4.7

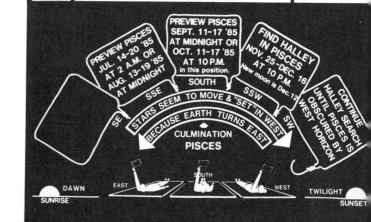

PHOTOGRAPHIC RECORD

PSC

Date

VISUAL RECORD

Film:		Type:			ISO		
					ISO		

CAMERA:			CLAMP/TRIPOD EXPOSURES, SECS.		PIGGYBACK/DRIVE EXPOSURES, MINS.		
LENS:		TAKEN					

DATE:	FILM#	SHOT#	AT HRS.	USE APPLICABLE COLUMN							
				10				1M			
					15				2M		
						20				4M	
							40				8M

DATE:	FILM#	SHOT#	HRS.								

DATE:	FILM#	SHOT#	HRS.								

DATE:	FILM#	SHOT#	HRS.								

RESULTS AND NOTES

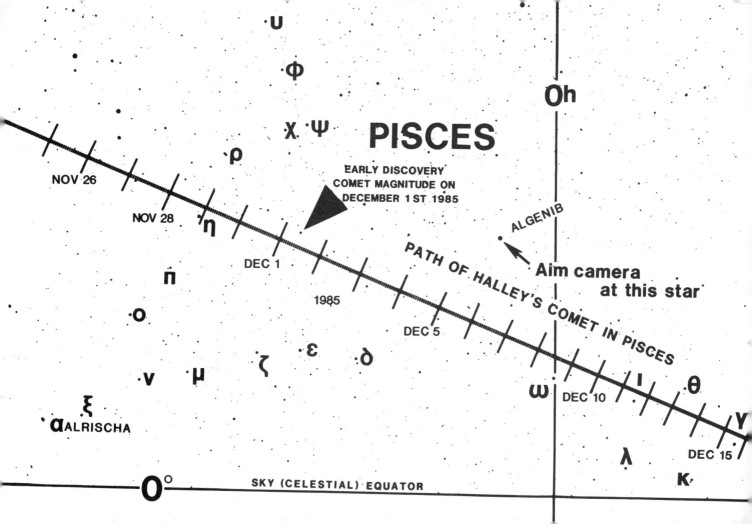

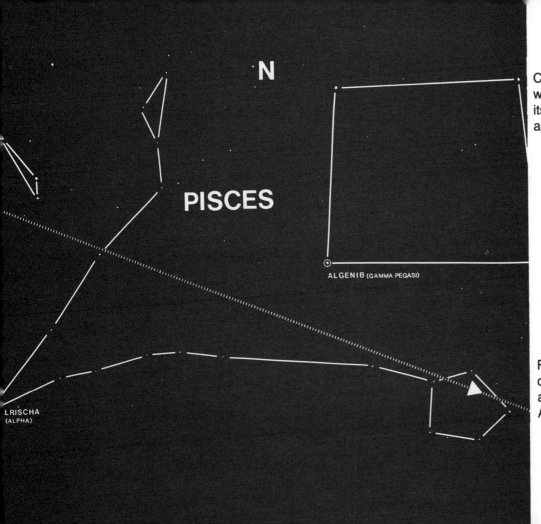

N

PISCES

ALGENIB (GAMMA PEGASI)

LRISCHA
(ALPHA)

Culmination dates and times: when the constellation will be at its highest, some 20° to 40° above the southern horizon.

PISCES

15 JUL.	5 A.M.
1 AUG.	4 A.M.
15 AUG.	3 A.M.
1 SEP.	2 A.M.
15 SEP.	1 A.M.
1 OCT.	MIDNIGHT
15 OCT.	11 P.M.
1 NOV.	10 P.M.
15 NOV.	9 P.M.
1 DEC.	8 P.M.
15 DEC.	7 P.M.
1 JAN.	6 P.M.

For Daylight Saving Time, add one hour to the times shown above during the months from April through October.

AQUARIUS

The brightest stars in Aquarius held an association with luck for ancient Arab astronomers, as the Arabic translations of the proper names of Alpha (α), Beta (β) and Gamma (γ) in Aquarius suggest. This area should prove lucky for us, too, because it is in this region that one can safely predict not only a first photographic recovery of Comet Halley but also first visual encounters as detailed in Chapter 8.

Fortunately the Water Bearer constellation, which has related to a heavenly sea since Babylonian days, offers excellent opportunities for extensive observation and the taking of test photographs. Begin taking photographs of the constellation itself during the new-moon period on August 16, 1985, when the kite-shaped grouping of the brightest stars will be in the highest culmination position at 1 A.M. At the next new moon the constellation will have passed "over the top" and will be very conveniently placed just to the right of the twelve-o'clock high position. Please try to observe Aquarius at every possible opportunity from September on. You can gain an important understanding of the motion of our earth from your observations. It will explain why Aquarius will be near setting in the west when the comet will pass through it in December '85–January '86 (see diagram, page 19). With your skymap you will be able to anticipate the path of the comet below the "kite" already in September 1985 and test the need for very gradually changing the angle of your camera, tilting it to the right, just like your starframe. As the constellation nears its setting position, you will want to outline (with your starframe or viewfinder) the identical area you first viewed and observed month after month.

In your September constellation photographs, taken near the culmination of Aquarius, try to be on the lookout for M2, the globular cluster of stars that appears with the brightness of a star of 7th magnitude. It will be a good object on which to try to improve your photographic skills.

In January 1986, when Comet Halley will soar along the kite between Sadalmelik and Sadalsud, it will approach 4th magnitude and should present no brightness problems to your camera. By this time also, the comet should show the first beginnings of a tail.

40

THE Waterbearer — AQR / Aquarii

L.Y.	BRIGHTEST STARS	MAG.
950	α ALPHA **SADALMELIK** LUCKSTAR OF THE KING (AR)	3.0
980	β BETA **SADALSUD** LUCKIEST OF LUCKSTARS (AR)	2.9
95	γ GAMMA **SADALACHBIA** LUCKSTAR OF TENTS (AR)	3.8
98	δ DELTA **SKAT** SHIN OF WATERBEARER (AR)	3.3
170	ε EPSILON **AL BALI** SWALLOWER'S LUCKSTAR (AR)	3.7
75	ζ ZETA **ZETA AQR**	3.6

PREVIEW AQUARIUS JULY 14-20 '85 AT MIDNIGHT

PREVIEW AQUARIUS SEPT. 11-17 '85 AT MIDNIGHT

PREVIEW AQUARIUS NOV. 9-15 '85 AT 10 P.M.

SEE HALLEY IN AQUARIUS DEC. 21-JAN.31 '86 AFTER SUNSET & EVEN. TWILIGHT

SOUTH

SSE — SSW

STARS SEEM TO MOVE & "SET" IN WEST

SE — SW

BECAUSE EARTH TURNS EAST

CULMINATION

AQUARIUS

DAWN / SUNRISE — EAST — SOUTH — WEST — TWILIGHT / SUNSET

PHOTOGRAPHIC RECORD

AQR

Date   **VISUAL RECORD**

Film:		Type:		ISO	
				ISO	

CAMERA:			CLAMP/TRIPOD EXPOSURES, SECS.	PIGGYBACK/DRIVE EXPOSURES, MINS.
LENS:		TAKEN		

DATE:		SHOT#	AT HRS.	USE APPLICABLE COLUMN								
	FILM#			10				1M				
					15				2M			
						20				4M		
							40				8M	

DATE:		SHOT#	HRS.									
	FILM#											

DATE:		SHOT#	HRS.									
	FILM#											

DATE:		SHOT#	HRS.									
	FILM#											

RESULTS AND NOTES

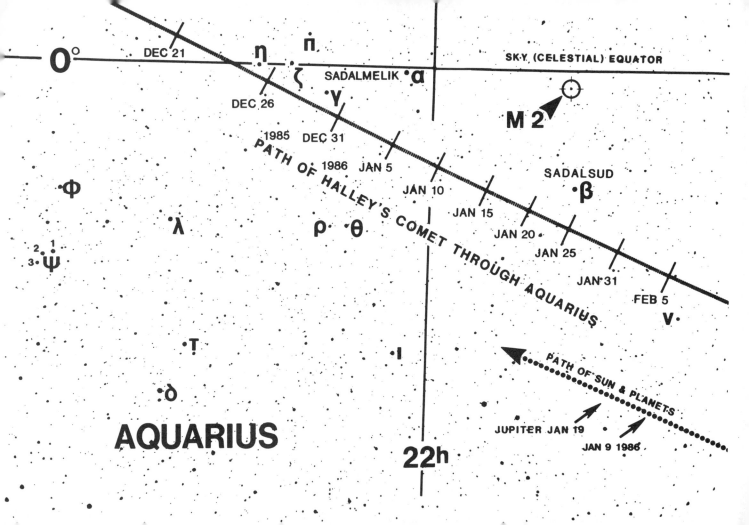

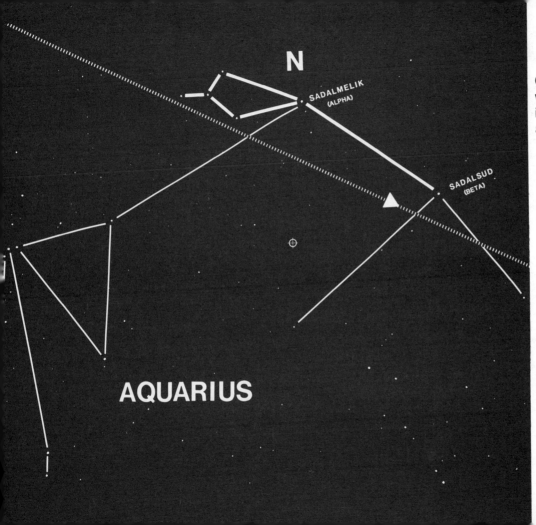

Culmination dates and times: when the constellation will be at its highest, some 20° to 40° above the southern horizon.

AQUARIUS

15 JUN.	5 A.M.
1 JUL.	4 A.M.
15 JUL.	3 A.M.
1 AUG.	2 A.M.
15 AUG.	1 A.M.
1 SEP.	MIDNIGHT
15 SEP.	11 P.M.
1 OCT.	10 P.M.
15 OCT.	9 P.M.
1 NOV.	8 P.M.
15 NOV.	7 P.M.
1 DEC.	6 P.M.

For Daylight Saving Time, add one hour to the times shown above during the months from April through October.

CAPRICORN

Having followed the comet into the Aquarius evening twilight to the vicinity of the sun, you will readily grasp the simple fact that as the comet nears the sun it will be in close solar proximity at *both sunset and sunrise*. In fact, "perihelion," the point when Comet Halley will be nearest to the sun, will occur on February 9. Unfortunately, the comet will then be lost in the sun's daylight glare. Toward the end of February the comet will reappear low in the southeastern sky *at dawn*. It will still be in the general vicinity of old Sol, which it will, by now, have passed. The brightness of the visitor will have increased to 2nd magnitude, and the tail should have curved and lengthened considerably. During the time the comet is in Capricorn, the constellation of the Sea Goat, our observations will have to be conducted during morning astronomical twilight.

Capricorn is not an easy constellation to observe, except near culmination, when its shape, reminiscent of a bikini bottom, can be recognized. Algedi, Alpha one *and* Alpha two ($\alpha^1 \alpha^2$), being the first stars in the constellation to rise in the east, will be stacked vertically just as the entire bikini tilts to the left in its rising position during the February dawn. The more you are able to familiarize yourself with the constellation in advance, especially as it relates to the brighter stars of Sagittarius (which will have risen two hours before Capricorn), the easier it will be for you to find the comet. Fortunately, the comet will now be much more conspicuous to your newly trained eye. You should have no trouble finding it toward the end of February, if not before.

THE Goat

CAP
Capricorni

L.Y.	BRIGHTEST STARS	MAG.
1,000	α_1 ALGEDI — THE GOAT (AR)	4.2
100	α_2 ALPHA	3.6
100	β DABIH — SLAUGHTERER'S STAR (AR) BETA	3.1
100	γ NASHIRA — GOOD TIDINGS STAR (AR) GAMMA	3.8
49	δ DENEB ALGEDI — TAIL OF THE GOAT (AR) DELTA	2.9

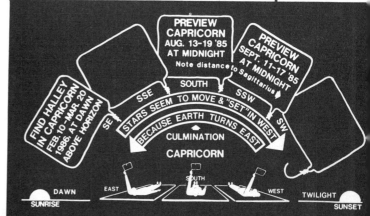

PREVIEW CAPRICORN AUG. 13–19 '85 AT MIDNIGHT
Note distance to Sagittarius

PREVIEW CAPRICORN SEPT. 11–17 '85 AT MIDNIGHT

FIND HALLEY IN CAPRICORN FEB 10–MAR. 20 1986, AT DAWN ABOVE HORIZON

STARS SEEM TO MOVE & "SET" IN WEST BECAUSE EARTH TURNS EAST

SE — SSE — SOUTH — SSW — SW

CULMINATION
CAPRICORN

DAWN — SUNRISE — EAST — SOUTH — WEST — TWILIGHT — SUNSET

PHOTOGRAPHIC RECORD	**CAP**	Date	**VISUAL RECORD**

Film: **Type:**   ISO / ISO

CAMERA:

LENS:			**TAKEN**	**CLAMP/TRIPOD** EXPOSURES, SECS.				**PIGGYBACK/DRIVE** EXPOSURES, MINS.			
DATE:	**FILM#**	SHOT#	AT HRS.	USE APPLICABLE COLUMN							
				10				1M			
					15				2M		
						20				4M	
							40				8M
DATE:	**FILM#**	SHOT#	HRS.								
DATE:	**FILM#**	SHOT#	HRS.								
DATE:	**FILM#**	SHOT#	HRS.								

RESULTS AND NOTES

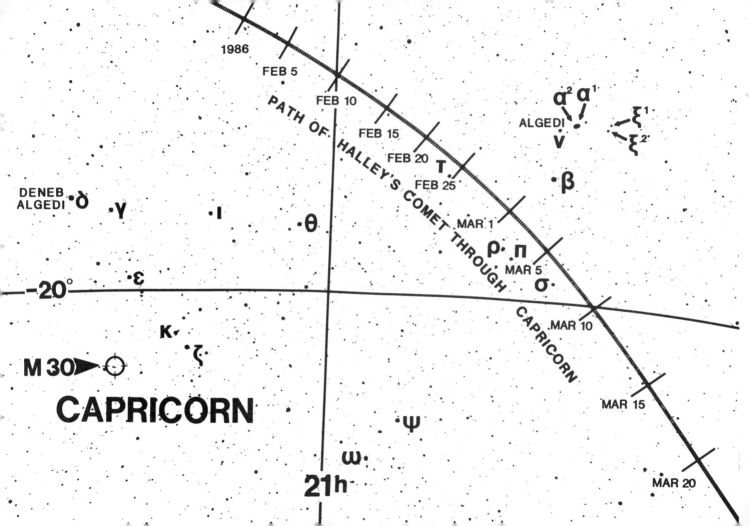

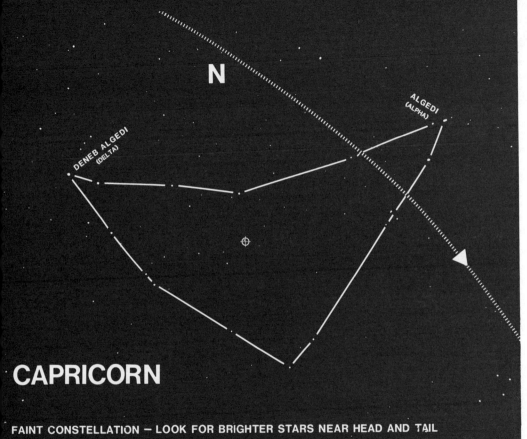

N

ALGEDI
(ALPHA)

DENEB ALGEDI
(DELTA)

CAPRICORN

FAINT CONSTELLATION – LOOK FOR BRIGHTER STARS NEAR HEAD AND TAIL

Culmination dates and times: when the constellation will be at its highest, some 20° to 40° above the southern horizon.

CAPRICORN	
15 MAY	5 A.M.
1 JUN.	4 A.M.
15 JUN.	3 A.M.
1 JUL.	2 A.M.
15 JUL.	1 A.M.
1 AUG.	MIDNIGHT
15 AUG.	11 P.M.
1 SEP.	10 P.M.
15 SEP.	9 P.M.
1 OCT.	8 P.M.
15 OCT.	7 P.M.
1 NOV.	6 P.M.

For Daylight Saving Time, add one hour to the times shown above during the months from April through October.

SOUTH
NORTH

SAGITTARIUS

Sagittarius harbors many treasures. When seen from a dark location it brings myriad stars to light near the center of our galaxy. Color exposures of a minute or more reveal the pink hydrogen glow of M8, the famous Lagoon nebula, one of the many exquisite telescopic objects in this rich region.

Sagittarius is where the very beauty of the night sky may challenge you to improve your photographic skills. It might even lure you enough to build a STELAS "stop-the-earth" system (see Chapter 12) for your camera which will vastly improve your pictures and your reach to faint magnitudes. Should you fall victim to the urge to get a telescope in 1985 (see "Resources," page 77), you will be able to use its motor or linkage for polar-aligned "piggyback" photography. The optics will reveal nebulae, clusters and even galaxies starting in the summer sky and leading you into autumn astronomical delights. Once you have sighted Comet Halley you will be able to view it telescopically for greater insights.

The "teapot" is the popular pictogram with which to find the Sagittarius starfield and Nunki, Sigma (σ) in Sagittarius, the brightest of the stars in the area. You can spot the teapot as early as June or July and begin to relate it visually to Capricorn following behind (to the left).

Try to get an understanding of the distance between the handle of the (Sagittarius) teapot and the waistline of the (Capricorn) bikini in autumn 1985. The tilt to the right of your camera in the fall of 1985 will be exactly the opposite of the tilt of your camera to the left in early 1986. You will soon sense the motion of our planet. Any motion in the night sky will be best understood when compared and related to the apparent similar motion of the sun in the daylight sky. Both Capricorn and Sagittarius, being southern constellations, will be easier to spot from California, Texas or Florida than from states farther north. When Comet Halley travels through Sagittarius, its tail will have grown to near-maximum length. Note that it will again point away from the sun, which will be to its left (east) and may still be below the horizon.

THE Archer — SGR — Sagittarii

L.Y.		BRIGHTEST STARS	MAG.
210	σ SIGMA	NUNKI (BABYLONIAN)	2.0
250	α ALPHA	RUKBAT — KNEE OF THE ARCHER (AR)	4.1
120	γ GAMMA	AL NASL — THE ARROWHEAD (AR)	3.0
82	δ DELTA	KAUS MERIDIONALIS — MIDDLE BOWSTAR	2.7
85	ε EPSILON	KAUS AUSTRALIS — SOUTHERN BOWSTAR	1.9
98	λ LAMBDA	KAUS BOREALIS — NORTHERN BOWSTAR	2.8
78	ζ ZETA	ASCELLA — ARMPIT (LATIN)	3.0

| PHOTOGRAPHIC RECORD | | | SGR | | Date | VISUAL RECORD |

PHOTOGRAPHIC RECORD   SGR   Date   VISUAL RECORD

Film:		Type:		ISO								
				ISO								

CAMERA: **CLAMP/TRIPOD**   **PIGGYBACK/DRIVE**

LENS: **TAKEN**   **EXPOSURES, SECS.**   **EXPOSURES, MINS.**

		SHOT#	AT HRS.	USE APPLICABLE COLUMN								
DATE:	FILM#			10				1M				
					15				2M			
						20				4M		
							40				8M	

		SHOT#	HRS.									
DATE:	FILM#											

		SHOT#	HRS.									
DATE:	FILM#											

		SHOT#	HRS.									
DATE:	FILM#											

RESULTS AND NOTES

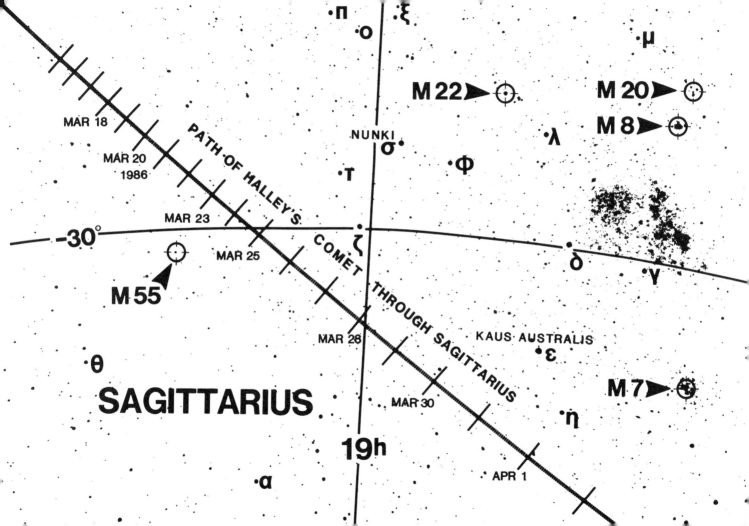

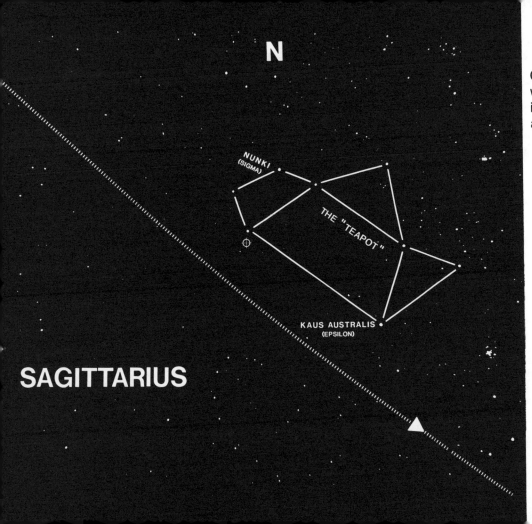

N

NUNKI
(SIGMA)

THE "TEAPOT"

KAUS AUSTRALIS
(EPSILON)

SAGITTARIUS

Culmination dates and times: when the constellation will be at its highest, some 20° to 40° above the southern horizon.

SAGITTARIUS	
15 APR.	5 A.M.
1 MAY	4 A.M.
15 MAY	3 A.M.
1 JUN.	2 A.M.
15 JUN.	1 A.M.
1 JUL.	MIDNIGHT
15 JUL.	11 P.M.
1 SEP.	10 P.M.
15 SEP.	9 P.M.
1 OCT.	8 P.M.
15 OCT.	7 P.M.
1 NOV.	6 P.M.

For Daylight Saving Time, add one hour to the times shown above during the months from April through October.

HYDRA AND BEYOND

April 1985 presents a useful opportunity to do the earliest possible practical work in connection with Halley's Comet. We can begin to study the area where Halley's Comet will make its exit from the vicinity of the earth and the sun one year hence, in 1986. The sooner we grasp the opportunity to fall in step with the heavens, the more productive our cometwatch will be.

After having vanished below southern horizons in early April 1986, Comet Halley will reemerge in evening skies in mid-April for us to view and photograph. Hydra the Water Serpent will be the first constellation in which the comet will become visible to Northern Hemisphere observers again. Soon after, it will cross farther north into the constellation of Crater the Cup, to reenter the convoluted Hydra before proceeding into Sextans, named after a navigational instrument also used to chart the heavens.

It is in this region that we can apply our previous Halley experiences to important new uses—a comet search in reverse. As we follow the fading comet and continue practicing our photography on the gradually dimming nucleus, we can test just how long it will be possible for us to keep it in sight. By April 20, 1986, the comet will still show up at about magnitude 4. About a month later it will have dimmed to 6th magnitude. Now we can test what the limit of our search capability will be as we prepare for a comet search of our own (see Chapter 13). We can improve our blinking technique because arriving and departing comets look, behave and photograph much the same.

It may well be that in going out to observe Hydra, Crater and Sextans in 1985, your eye will be drawn to the zodiacal constellations of Virgo and Leo to the north, or to the old and familiar Big Dipper, which will also be in the April skies.

We have shown only seven windows to the magnificent skies, each one an endless sandbox full of stars. Even as Comet Halley will leave us in 1986, not to return for seventy-six years, new discoveries beckon. You may want to heed the call of comets waiting to be discovered, starting in 1985. The magic of astronomy can make any of the celestial constellations into comet host constellations. Anywhere in the evening night or predawn sky, a comet may even now be approaching to leave its imprint on a film in your camera. It will reveal itself through simple comparison with an earlier photograph of the same sky region.

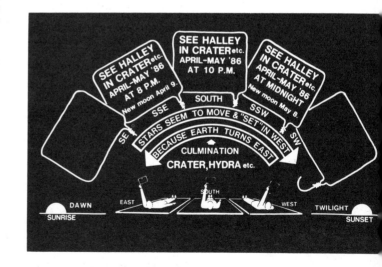

PHOTOGRAPHIC RECORD

Date | **VISUAL RECORD**

Film:		Type:		ISO		
				ISO		

CAMERA: | CLAMP/TRIPOD EXPOSURES, SECS. | PIGGYBACK/DRIVE EXPOSURES, MINS.

LENS: | **TAKEN**

	SHOT#	AT HRS.	USE APPLICABLE COLUMN								
DATE: **FILM#**			10				1M				
				15				2M			
					20				4M		
						40				8M	

	SHOT#	HRS.									
DATE: **FILM#**											

	SHOT#	HRS.									
DATE: **FILM#**											

	SHOT#	HRS.									
DATE: **FILM#**											

RESULTS AND NOTES

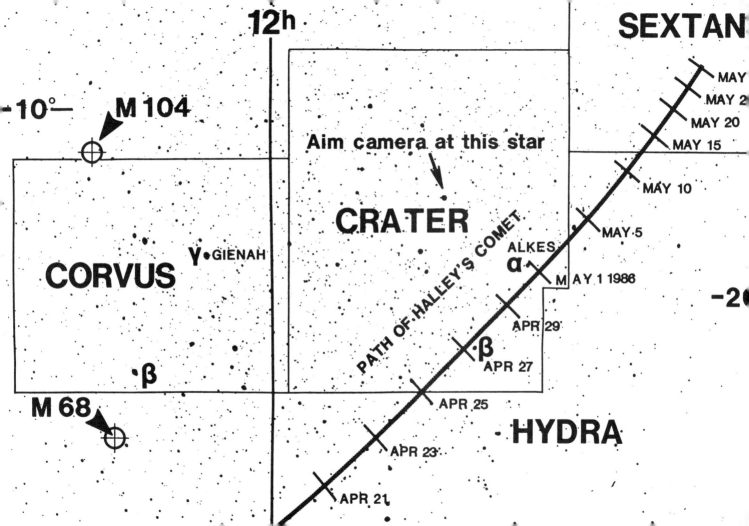

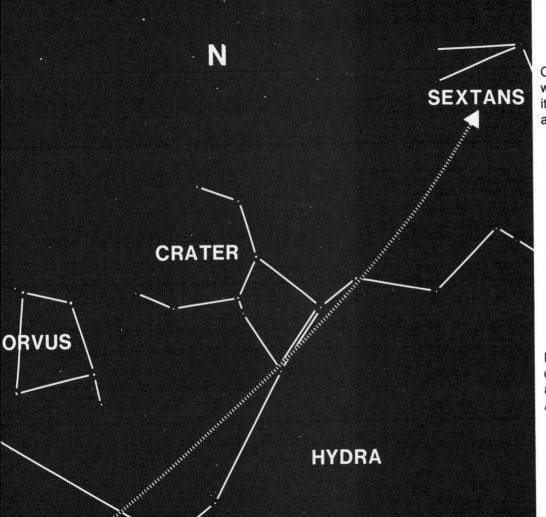

N

SEXTANS

CRATER

ORVUS

HYDRA

Culmination dates and times: when the constellation will be at its highest, some 20° to 40° above the southern horizon.

HYDRA

1 JAN.	5 A.M.
15 JAN.	4 A.M.
1 FEB.	3 A.M.
15 FEB.	2 A.M.
1 MAR.	1 A.M.
15 MAR.	MIDNIGHT
1 APR.	11 P.M.
15 APR.	10 P.M.
1 MAY	9 P.M.
15 MAY	8 P.M.
1 JUN.	7 P.M.
15 JUN.	6 P.M.

For Daylight Saving Time, add one hour to the times shown above during the months from April through October.

55

8. First Encounters: January 1986

For the earliest possible visual sighting, buy, borrow or swap for a pair of binoculars. If the days are clear, a first trip to a higher and darker location between January 5 and 10, 1986 should be well worth your time. The comet should be visible early in the first darkness of the evening, so even youngsters can be taken along for an event they may never forget.

Keep an eye on satellite weather pictures that television stations show with news broadcasts. Obviously, you will not be able to see the comet if *both* the view out your window and the satellite records on TV show solid cloud cover. But such conditions are known to change quite rapidly. Forecasters are often caught with their raincoats on when they might have left them in the closet. That is why the sequential pictures taken from high altitudes are so valuable. They can reveal a clear patch moving to your area at about the time you may plan your Halley excursion. Think positively. It will be easier to maintain good thoughts if you have already taken early Halley photographs in December.

Still, catching first sight of the comet steeped in history will be a very special event even for experienced stargazers.

You may want to select an early-January 1986 Halley-viewing date, not only because of favorable lunar positions with no moon in sight at that time, but also because it is during these days that the comet first will reach magnitudes somewhere between 5.2 and 4.8, well within sight of the naked eye *under dark skies.* You may have selected your observing site long before now. Selecting a location from which you can spot the two stars south of Sadalmelik, Alpha (α) in Aquarius—the ones with brightnesses of 5.3 and

4.7 respectively—can involve a pleasant evening drive in summer or autumn of 1985. Finding a good observing site is critically important and should be done well in advance, during the fall of 1985.

Here are some of the items you should take with you on all your star expeditions: extra blankets, even though you will dress warmly; *Halley's Comet Finder* with its log pages and a pencil for entries; a compass to establish your north–south axis and for finding the southwest position; a red-masked flashlight. (Buy a flashlight and some nail polish, and paint the bulb red.) Bring some binoculars if you can. If you have an SLR 35mm. camera, bring it, along with clamp or tripod and any accessories you may have built for it. Be sure the camera is loaded with the fastest color film you can find (see Chapter 11). Keep the camera warm or in a car until just before use, so that the lens will not fog over too soon.

In January 1986 you should be in position to start your observations not later than an hour after the sun has gone down and you have seen it disappear below the horizon. The comet should be just above the point where you saw the sun set. One hour after sunset, the sun will be 12 degrees below the horizon and it will be getting quite dark. Still, not until 90 minutes after sundown will the sun be 18 degrees below the horizon, after which it will be as dark as it can get. Use the interval to establish your bearings, north behind you, south straight ahead. You can also set up your tripod or makeshift camera support. During this time your eyes will adapt to the dark quite naturally. Use only your red flashlight to refer to notes or to check your compass. It soon should be time to look for Halley's Comet.

If you have a starframe, tilt it to the right as shown. Be sure the bottom corner is aimed in a southwest direction. Now start aligning the starframe with the Kite in Aquarius, which should be readily visible. The only possible variable factor in this routine has to do with your position on the surface of our earth. If you live in southern latitudes, say in Arizona, Louisiana or Georgia, you will hold your *tilted* wire frame one starframe height (8 inches) above your horizon. Observers farther north would hold their finder frame a little lower, about 20 degrees above the southwest point on the horizon.

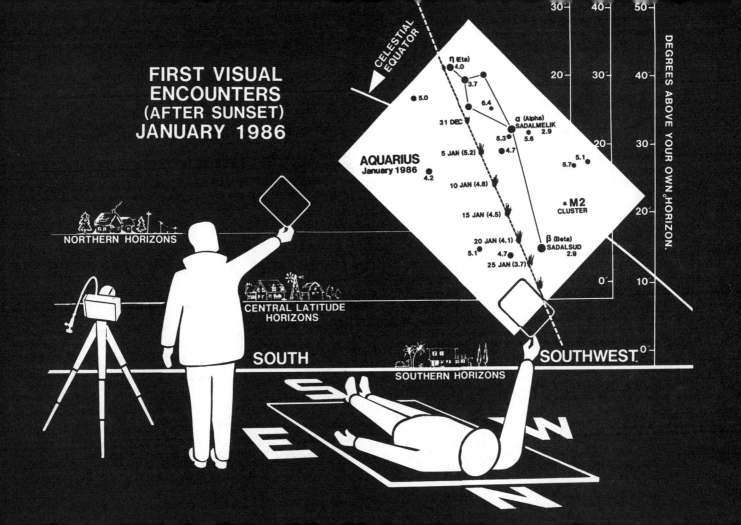

FIRST VISUAL
ENCOUNTERS
(AFTER SUNSET)
JANUARY 1986

CELESTIAL EQUATOR

η (Eta)
4.0

3.7

5.0

6.4

31 DEC

α (Alpha)
SADALMELIK
2.9

5.3

5.6

AQUARIUS
January 1986

5 JAN (5.2)

4.7

5.1

5.7

4.2

10 JAN (4.8)

15 JAN (4.5)

♦ M2
CLUSTER

20 JAN (4.1)

β (Beta)
SADALSUD
2.9

5.1

4.7

25 JAN (3.7)

30   40   50

20   30   40

DEGREES ABOVE YOUR OWN HORIZON.

30

20

10

0

NORTHERN HORIZONS

CENTRAL LATITUDE
HORIZONS

SOUTH

SOUTHERN HORIZONS

SOUTHWEST

Cometwatchers on latitudes farther north will hold their frames even closer to the southwestern horizon. If you have forgotten your starframe, search the area just above the point where you observed the sun to disappear below the horizon.

Tripod-mounted cameras should be similarly pointed. This is a fine time for photography. About six exposures will be a good investment on such a first session. Count out 5 seconds slowly, then 10, 15, 20, 25 and 30. Shoot with the aperture f/stop wide open. Be sure you take notes after each exposure or, in the case of the longer exposures, while the shutter is held open. For cameras tracking the sky, try 1-minute, then 2-, 3-, 4-, 5- and 6-minute exposures. You may even want to shoot some devil-may-care exposures of 8 or 10 minutes. Do not forget to tilt your camera 45 degrees *to the right* so that your photos will match the ones in this book.

Time will fly, and soon you will see Sadalsud, Beta (β) in Aquarius, nearing its setting position. It will be just to the left of this brightest star in Aquarius that Comet Halley too will set at the end of January. So there will be no time to lose as we search for the comet. Remember, if you do not spot Halley on this, your first outing, you may still bring a picture of the comet home in your camera, an unexpected treasure which, properly enlarged and framed, can hang in your den for years to come. It will serve as a treasured reminder of the night, perhaps two or three days after your first outing, when you suddenly spotted Comet Halley with the naked eye.

There it is! The visitor from the past paying its respects on the way into the future. You will wonder how you could have missed it before, because it is so strikingly different from anything you have ever seen in the night sky. There is even a small tail pointing away from the sun which set here so recently. You watch it for as long as you can, perhaps a precious half hour. As you head back you know that the comet is now heading for its rendezvous with the sun, not to be seen again until late in February 1986. You also know that you will rise early at the end of February for your second encounter with history.

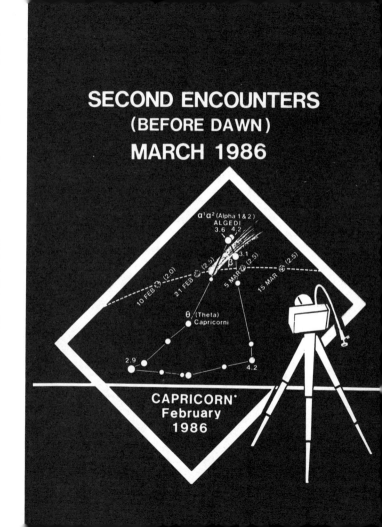

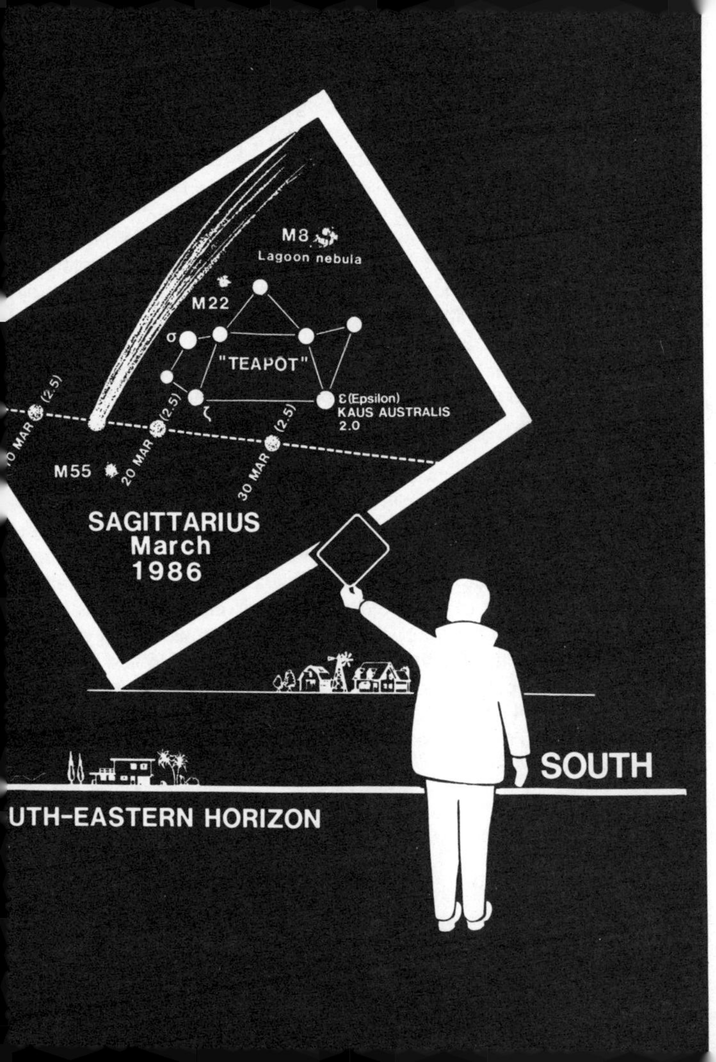

M8
Lagoon nebula

M22

σ

"TEAPOT"

ε (Epsilon)
KAUS AUSTRALIS
2.0

30 MAR (2.5)

20 MAR (2.5)

30 MAR (2.5)

20 MAR (2.5)

M55

SAGITTARIUS
March
1986

SOUTH

UTH-EASTERN HORIZON

9. Second Encounters: March 1986

Most people do not enjoy getting out of bed in the morning. Love for the stars may change all that. So will Halley's Comet fever.

You may not know what you usually miss by not seeing the first dawning of a new day. An eerie first glow may precede it. This is the "false dawn" talked about by the poet Omar Khayám—what astronomers call the "zodiacal light." It is caused by reflections from interplanetary dust of the light of a sun still far below the horizon. The shape in which this light is concentrated matches that of the "haystacks" in Chapter 14.

The stars one sees in the eastern morning sky belong to constellations that will not culminate for another three months. In a way they present a very private celestial preview of joys to come. Private these viewings almost always are, because habit has decreed that time around dawn be spent sleeping. Even the most dedicated night owls seem to turn in at about this time, missing the thrill of the arrival of the new day.

The song of an early bird is heard; perhaps the sound of a distant bell which may chime every hour but is drowned out by noise at all other times. In rural areas there will be the crowing of a cock and the response of another, far, far, away The sky begins to glow in hues never seen before, heralding the sunrise. Wispy clouds may stretch along the eastern horizon in delicate layers. Above us, the brighter stars still shine on an ashen background. Only the west still seems to cling to darkness, loath to surrender the night.

It is into such romantic skies that Comet Halley should make its second entry for northern observers in later February and early March 1986. The earlier we spot the comet, the closer to the sun it will be, having just rounded it at perihelion. Only very gradually, as mid-March approaches, will the mysterious snowball take shelter in the early-morning darkness, gradually to become a night visitor again. One almost expects comets to be nocturnal along with stars and meteors, but, as we know now, they also inhabit the daylight skies, unseen.

You may want to plan a "surprise sunrise comet party" on the weekend of March 15. You may be surprised by suddenly finding the comet, and then again you may be surprised at merely witnessing a magnificent sunrise. Another surprise may be a sudden change in the weather shrouding the eastern horizon in unexpected fog. But now is the time when early risers or late revelers may be able to find the spark of the comet in the glow of the flames of the approaching sunrise.

Halley watchers in southern latitudes will have better luck. Those in central or more northern latitudes may need to wait a little longer, until, in mid-March, the comet may reveal itself to them. For this second recovery, though, observers north of an imaginary San Francisco–Philadelphia line may have increasing difficulties in finding the object. An unbroken clear view of the southeastern horizon and clear dark skies become mandatory. Take photographs; your camera may reveal what your eye cannot yet see.

Even though the Capricorn skymap shows the February–March path of Halley's Comet, you may soon wish to refer to the overlapping Sagittarius starmap, if only because it centers on the brighter stars of the "teapot." These will have been in the predawn sky for about an hour and will be easier to spot than the rising Sea Goat. Above all, you must remember to tilt camera and starframe 45 degrees *to the left* for early-morning patrols (just as you tilted the Aquarius "kite" frame to the right for the evening post-sunset observations.) The bottom of the teapot itself should be about parallel to the horizon at this stage. It will help if you have seen Sagittarius and its distance and relationship to Capricorn in autumn 1985.

Although you may want to go and try your luck daily around these dates, the best weekend with good moonless mornings presents itself on the 15th of March 1986. Some nearby skiing resort may be just the spot. Prepare yourself for a treat when the comet comes in sight. The head by now will have brightened to 2nd or even 1st magnitude, and the tail will have grown to span the width of your starframe. It will always point away from the sun. Take too many rather than too few pictures while the comet is in Sagittarius. Shoot the works. "The Teapot and the Comet": you may take a memorable picture. Try every best exposure time you ever used and "bracket" it by taking another pair, one a little shorter, the other a little longer. If in doubt, try 10-, 12-, 15-, 18-, 20-, 24- and 30-second exposures. You cannot take too many pictures now. In this fashion, too, you may be able to offer an original photo to your neighborhood newspaper. If it is a fine record, they might print it.

Just as it rose above the southeastern horizon in early March, Comet Halley will in early April 1986 swing back down to where it becomes impossible to observe from northern sites. As it gets brighter and the tail longer, the comet will dazzle those who live in the Southern Hemisphere. Many of the comet-struck will journey to South America, even Australia, to get front-row-center seats for the spectacle of a lifetime. The tail will be at its longest, one starframe wide, and no horizons will block the view of the celestial voyager as it travels high in the Southern Hemisphere autumn night sky. Lucky persons will book reservations for the first moonless weeks of April 1986. Lima, Peru, will be fine; Rio de Janeiro in Brazil will be even better. Treat yourself, even if you never take another vacation. This is it, unless you can wait until the year 2061.

10. Third Encounters: April 1986

In mid-April 1986 Comet Halley comes into view once more for Northern Hemisphere cometwatchers. Happily, this time it will come to stay above our southern horizon, at first for a brief month of naked-eye observation and then for as long as optical aids and cameras can keep it in sight. Not until August 1986, when the visitor will reenter the sun-drenched daylight sky, will it finally be beyond the reach of most amateur instruments.

As it returns to the deepest reaches of the solar system the comet will get gradually fainter. We can track it on its outward journey and gather much valuable additional observing experience.

Let us not regard the third encounters as a Halley's Comet finale, but rather as a beginning of a vigil for many other comets to come. By now we will have assembled much firsthand experience concerning the heavens, several of the constellations and some principal stars. We may even have taken our first successful astrophotographs. All our new knowledge can be put to immediate and productive use. Many wonderful opportunities present themselves for amateurs to make meaningful contributions to science and to the astronomical community.

Just our ability to locate several specific areas in the evening, midnight or morning skies qualifies us with new and useful understanding. It will be invaluable when applied to other constellations. It will help us detect comets found just recently by others. The better-known periodic comets discovered over the last two hundred years can be accessible to us. We may even be

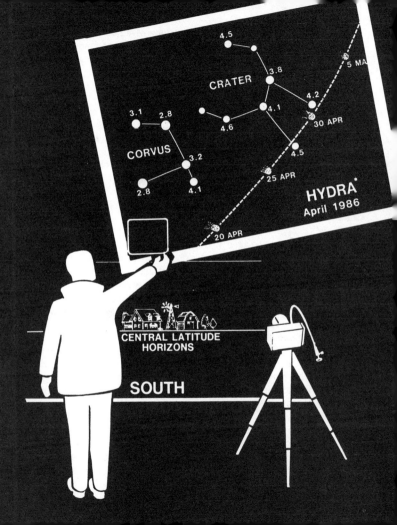

tempted to explore variable stars that may have revealed themselves during the blinking of our photographs. This is an area where amateurs worldwide are performing valuable work for professionals. For additional information contact the American Association of Variable Star Observers, 187 Concord Avenue, Cambridge, MA 02138, U.S.A.

Our third-encounter search takes place in the late evening at about 10 to 11 P.M. It will be centered on southern constellations. The comet will just have left Centaurus, the region of the heavenly Centaur, and will have entered Hydra the Water Serpent. We will be using the bright grouping of stars in the adjacent constellation of Corvus the Raven to align our starframe. Beta (β) and Gamma (γ) in Corvus are the brightest stars in the region, which will guide us even without a coathanger. They will help us find our way even after Halley's Comet will have crossed from the constellation of Hydra into Crater the Cup.

We should have ample time to search for the comet on this leg of its journey, although the moon will be waxing from first quarter to full between April 17 and 24 and will make observation difficult at the start. The weekend of April 26 should afford a little time before the waning moon rises in the east, to try to spot the comet in the constellation Crater. As the week progresses and the fading moon rises later and later, conditions will improve drastically even though the tail of the comet and the brightness of its head will have begun to shrink. Weather permitting, the weekend of May 3 may well be a good time to load loved ones, family and friends into a bus or a car and to head for the mountains. The comet, now outward bound back to the deep freeze of space, will be visible all night and will offer some good chances for photography. The sky will be dark until the early-morning rising of a crescent moon.

It will be easy for us to orient ourselves for these encounters. We should be heading and looking almost straight south at the culmination times given for the skymaps on pages 54 and 55. The starframe should be held in its horizontal culminating position or tilted slightly to the left or to the right as the alignment stars will suggest.

For April, May and even June 1986 observations, you may want to explore more than just the comet. Here is an opportunity for all to experience what amateur-astronomer "star parties" are about. Call your nearest planetarium or the earth-sciences department at a college in your area, or contact the department of astronomy at some nearby university. They will know where the starlovers meet. Then telephone a member of such an amateur-astronomy club and arrange to join an observing session. You can be quite sure that many a telescope will be aimed at Halley's Comet for weeks, even months, to come. Although the comet will become fainter with the passing weeks, you may get to see it more clearly than you have done before. The memories of the event will be indelibly reinforced in the mind of some youngster for whom this may be a first telescopic experience and a last chance to view Comet Halley—until the year 2061. For you too, a possibly emotional last visual leavetaking may become a night to remember.

After looking at the departing comet and perhaps a less famous approaching one through several instruments, you may decide to take your next step toward the starry skies. A medium-priced telescope might catch your fancy, or perhaps a simple spyglass type. It may even be a home-built telescope through which a teenager has let you observe the tail structure of the comet. He or she may tell you many things about Halley, but, with special pride s/he will recount for you that s/he actually ground and polished the mirror for the instrument, the surface that reflected the cometlight to your eye.

Don't resist the lure of the stars, the rapture of deep space. Not only do astronomy clubs welcome new members, but they know the best times and places for your observations. You can learn how to construct your own giant Dobsonian telescope out of a cement-forming tube. Professional guest speakers will challenge you at meetings. You will want to learn of ongoing search projects that can put you on the threshold of genuine discoveries.

Such a star party can be a wonderful beginning. You may yet be out there for the rest of 1986, with your own telescope following Halley's Comet beyond Mars, even as you patrol the evening and morning "haystacks" to find your own comet, one that will forever bear your name (see Chapter 14).

11. Collecting Cometlight: Simple Astrophotography

It seems very unlikely that Halley's Comet photographs have ever been taken on roll film as we know it today; certainly it was never recorded on color roll film. All the wonderful developments in the field of photography had not been invented in 1910 when Halley's Comet last came by. The treasured few images we have were taken on glass plates at principal astronomical observatories. All are very valuable today, being the very first actual pictures of the historic comet.

During the 1985–86 apparition anyone with a 35mm. SLR (single-lens reflex) camera can photograph the comet and keep records for the future. Cameras need only have a "B" (bulb) setting, so that they can be held open with a cable release during the exposures and then closed again. Lenses should be any standard type. Old cameras or even secondhand models will perform fine service as long as the above requirements can be met. Consider buying a used camera in a photo store or even a pawnshop.

The arrival in the marketplace of exceptionally "fast," i.e. sensitive, films could not have been better timed. The new color films designed for use under very low light conditions are ideally suited for our purposes. Since we are going to aim at the night sky, and no fast-moving objects will be involved, we can and will use longer exposures than usual, in fact much longer. For this reason the camera itself must not move. It should be held quite still, attached firmly to a tripod or an inexpensive clamp (see figure at right).

If the recommendations for visual observing of the various encounters are followed, our photo targets will be easy to find. We will always be aiming the camera in the same directions *and at the same angles* as our starframes. In

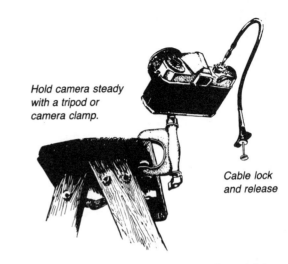

Hold camera steady with a tripod or camera clamp.

Cable lock and release

our early-recovery skymaps of Taurus, Aries and Pisces bright stars are marked; they should be centered in the viewfinder each time we photograph these regions. In this manner we will build a collection of shots with identical fields of view. Pairs of such photographs are used to find the comet while it is still faint, and only our camera may spot it. These matching sets of slides will reveal the positions of the new arrival by blinking on and off or back and forth in our discovery boxes or machines (see Chapter 13).

Start by photographing page 67 right onto your film. Take two exposures of this explanatory notice, completely filling the viewfinder so that the developing lab will not think that your night film is unexposed. Do not forget to refocus your camera after these shots to infinity (∞) and return the exposure dial to the "B" setting. For processing you may wish to enclose a written note with your films explaining that you are conducting a Halley's Comet program; anything to help the lab technicians and to avoid having your precious photographs returned cut or unmounted.

As with all scientific projects, you will want to practice your techniques as far in advance as possible to obtain your own firsthand information on what is best for your location(s) and your camera. You may wish to try out several of the high-speed films and note the differences between them. You may want to compare film processors for service and speed, even courtesy. The preceding chapters are all aimed at such preparations. They combine maximum chances to observe the Halley host constellations with early photographic opportunities. You can plan and work on your project many months in advance. Your end results will reward you.

In case you have no time to prepare, here is what you do: Ask for and load the *fastest* color film in your camera (see page 24). Transparency slide film offers the greatest versatility for our purposes. You can always have prints made, like the starmaps in this book, after having "blinked" and compared the slides.

SHOOT AT INFINITY ◀

Focusing Ring.

ft 4 5 7 10 15 30 ∞
m 1.2 1.5 2 3 5 10

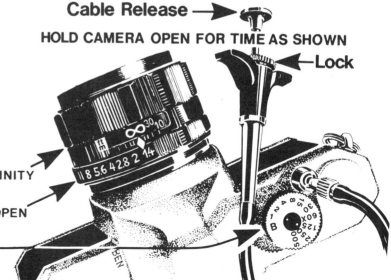

Cable Release ➡

HOLD CAMERA OPEN FOR TIME AS SHOWN

◀—**Lock**

Focusing Ring: ALWAYS AT INFINITY

Aperture Ring: ALWAYS WIDE OPEN

Exposure Dial: ALWAYS AT **B**

Enter your film information on your record page 67 in soft pencil so that you can erase it and reuse it. In this manner your record page becomes an integral part of your film. Now attach your camera securely to clamp or tripod and aim it at the sky on the dates and at the times listed. Double-check your target area with a starframe which you can hold behind the camera to verify your field *and its angle.*

Focusing presents no problems: *always* set your lens at infinity (∞). Aperture (lens opening) is easy: *always* use your lens at its widest possible setting, the f/1.4, 1.8 or 2 opening.

Exposure time is the only variable with which you will experiment. It offers many choices and it controls how much starlight enters your camera to accumulate on the film.

Make your first exposure 10 seconds long. Increase the next to 15 seconds. Then shoot one at 20 seconds. Make the last one an exposure of 30 seconds. You may also wish to test what happens at 12 or 18 seconds or at 1 minute. Whatever else you may do, write down each exposure in the log provided in this book, because you will not remember your data on the following day, certainly not when the film comes back from the lab. It is then that you will prize your own notes, which will be your very best teacher on how to proceed in the future. You will find that the shorter exposures showed only the brightest stars, while the longer ones penetrated to fainter magnitudes. As exposures became longer than 20 seconds or so, the stars began to "trail," and instead of crisp round images you started to get short "star trails" revealing the revolution of our planet.

If your work will be done with the camera fixed to a tripod or a clamp, you will soon find the optimum exposure times for such photography. It may be a good idea to standardize your procedures once you have established the maximum time for which you can leave your shutter open before the stars begin to "trail." You will soon find that your best results will come from exposures of 15 to 20 seconds with a standard 50 mm. lens.

LENS WIDE OPEN ▶

Aperture Ring.

1.4 2 2.8 4 5.6 8 11 16 22

On the other hand, in the event that you will be able to build a "stop-the-earth" drive system (see Chapter 12) or have access to a telescope with a motor drive on which you can "piggyback" your camera, much longer exposures are at your disposal. The sky truly becomes the limit. Now the motion of our planet, as it revolves on its axis once every twenty-four hours, will be stopped by a motor that turns the telescope or camera in the opposite direction, also at a rate of one RPD (revolution per day). Stars will cease to record as trails in longer exposures and will—if the system is properly aligned—remain neat round images. Exposures will be limited only by the darkness of your sky: it will become possible to leave the camera open for much longer times. Now you should begin to experiment with *1-minute* exposures, double to *2 minutes,* redouble to 4, even try 8-minute records. Such photographs can collect the faint light of 8th-magnitude stars and reveal the comet at the earliest possible stages. A recovery in Taurus in November 1985 becomes an attainable goal. Should you try even longer exposures, you will find that eventually "sky fog" will begin to collect on your films. It is caused by very low levels of light which are in the night sky at all times, even under the darkest of conditions. It will make the background of all your photographs appear washed out and gray. The nearer you are to city skies, the sooner this is likely to happen. Again, test your own best results and follow them, once you have found the best exposures for your location.

Even though photography and visual observations are best conducted from the darkest possible locations, you may wish to see what kind of results you may achieve under the most adverse of city skies. By all means, try to check what various exposures will yield from the roof of your city dwelling or your suburban backyard. You will be surprised, especially when you try black-and-white film with certain filters. Black-and-white film has long been used for city night photography. Load your camera with Kodak Tri-X film. Then put a No. 25 red Wratten filter (available in any photo store) in front of your lens. Follow the exposure times as listed above, depending on whether you work with a fixed or a motorized mount. Keep accurate notes. You should just have such film developed but not printed (lowest-cost method). Then buy cardboard or plastic 2-inch-by-2-inch slide mounts to cut and frame the negatives yourself. When making such slides, be sure to write all the camera data, exposure and site information on the margins. Just as with color slides, you should always note the number of the particular roll of film before identifying each picture with its individual frame number. Such data should correspond to your log entries. (Example: Film #4, frame 1; Film #4, frame 2, etc.) Be sure to note which side of the film is "south."

While in any camera store, you may want to inquire about developing your own black-and-white negative film. It is easy and can be done at home. You need only a dark space briefly to transfer your film from the cassette into a small lightproof developing tank. The short two-step procedure can be performed in daylight. There is nothing to it. Above all, it allows you to view your pictures within the hour of taking them. Today, even color films can be developed at home. Check it out.

One final word on film speeds as concerns film processing. If you shoot under dark skies you may want to try having your film "pushed" during development. What this means is that ISO 800 film would behave like ISO 1600 film, ISO 1000 would act like ISO 2000 and so forth. Shoot two identical rolls of the same type of film and compare exposures of the same length "pushed" and "unpushed." In the case of fixed-tripod photographs you will record much fainter stars near the 15–20-second exposure limit, just before the stars begin to trail. It is well worth the test. All that may happen is that the film will look a little "grainier." You may not even notice that. It is the first smidgen of cometlight we are after. Once you have found the comet you can shoot it over and over again, perhaps on different types of film. You will treasure your discovery photograph the most. But then again, a nice picture of comet and sweeping tail in magnificent color will also be a fine reward, a treasure to be prized.

Edmund Halley would have envied you.

FILM PROCESSING TECHNICIAN PLEASE NOTE!

THE FOLLOWING ARE PART OF A

COMET HALLEY PATROL

PLEASE MOUNT ALL SLIDES.

FILM #:_____ FILM TYPE:_____ I.S.O:_____

DATE: __/__/____

12. Stop the Earth: More Astrophotography

When one finds out how important it is to be able to lengthen exposure times in order to record fainter stars, the 15–20-second time limit looms ever larger. A motor-driven telescope on which to mount a camera may be as far beyond reach as the stars themselves. But do not despair. If we are lucky enough to own an SLR camera as described in Chapter 11 we are able to start our quest. All we have to do is stop the earth so that the stars stand still and we can capture their light and the glow of the comet in long exposures. This chapter will tell how we can arrest the motion of the earth for very little money.

You will not need a lightmeter. Such automatic features do not work at night, and we will never use flash attachments. All we need for our purposes is to see a sharp image of a distant horizon through the viewfinder at the infinity setting (∞) and the ability to open and close the lens with a cable release on the "B" (bulb) or "T" (time) setting. Cameras are accumulators of starlight. Their lenses are all the "optics" that starlovers will need. Beyond that it would be ideal if we could make the stars stand still as long as possible to record them on film.

Cometwatchers are a breed apart. Nothing will stand in the way of their quest. Stopping the world and thus the stars is the least of their challenges when the rewards can be so high and discovery can bring a link with history.

We know that the earth rotates on its axis once every twenty-four hours. It stands to reason that any drive mechanism designed to turn a camera aimed at the stars in the opposite direction to that in which earth is turning, at the same slow rate of one revolution per day/night, will nullify the apparent stellar motion. Thus celestial objects can be made to stand still while they are being observed or photographed. In this manner even faint starlight has a chance to accumulate on just one spot of the film over a period of time. This will allow us to penetrate to much fainter magnitudes, because, quite literally, the motion of the earth will no longer pull the film out from under the cometlight or starlight.

Small motors with 1/24 RPH are available in the marketplace, and we recommend a reasonably priced "Synchron type" available from Celestron International (see Resources, page 77) as noted on the diagram (item B). The tiny 1/8-inch-diameter driveshaft when mounted parallel to the axis of our earth is all that will be required to "stop the earth" so that we can take our photographs more leisurely. Once we mount a "ball head" (item J) on the axle of our motor we can swing our photo system anywhere in the sky. Now we are ready to try much longer exposures.

Even the most preliminary tests will show the tremendous differences in results that can be achieved by doubling the 15-second exposure and making it 30 seconds long. Once you try doubling that to arrive at a 1-minute exposure you are entering the big time. Your photographs will not only surprise you, they will defy belief among your friends.

Even though you can attempt long exposures from city locations, especially when you try the red-filtering method described earlier (see page 66) in combination with black-and-white film, you will want to seek out the darkest possible skies to get the most out of your new capabilities. In such locations 115-volt current to power your drive motor may not be available. You can carry your own power supply with you in the form of a 12-volt automobile battery in combination with a 115–12-volt converter. Eighth-magnitude stars now are well within your grasp.

A word of caution: The "STELAS" (Stop The Earth, Lock All Stars) drive platform has been tested and works extremely well. But, as can be surmised from looking at the picture, a considerable load is being brought to bear on the small driveshaft of the Synchron motor because of the weight of the camera. The leverage through the ball head can support a *standard* camera and lens. *Do not attempt to use heavier telelenses.* "Long" telelenses are "slow" in any case. Too much light is lost in them, requiring extra-long exposures. Such lenses share the "light loss" of telescopes.

How to build a STELAS.

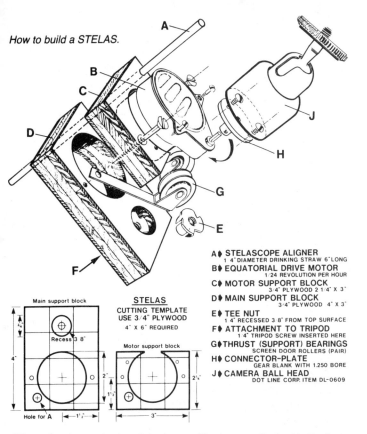

STELAS
CUTTING TEMPLATE
USE 3/4" PLYWOOD
4" X 6" REQUIRED

A⬥ STELASCOPE ALIGNER
1 4" DIAMETER DRINKING STRAW 6" LONG
B⬥ EQUATORIAL DRIVE MOTOR
1/24 REVOLUTION PER HOUR
C⬥ MOTOR SUPPORT BLOCK
3/4" PLYWOOD 2 1/4" X 3"
D⬥ MAIN SUPPORT BLOCK
3/4" PLYWOOD 4" X 3"
E⬥ TEE NUT
1 4" RECESSED 3 8" FROM TOP SURFACE
F⬥ ATTACHMENT TO TRIPOD
1 4" TRIPOD SCREW INSERTED HERE
G⬥ THRUST (SUPPORT) BEARINGS
SCREEN DOOR ROLLERS (PAIR)
H⬥ CONNECTOR-PLATE
GEAR BLANK WITH 1.250 BORE
J⬥ CAMERA BALL HEAD
DOT LINE CORP. ITEM DL-0609

STELAS on tripod with camera in place. The drinking straw always stays pointed at the North Star.

Often advertisements show telescopes with cameras attached to the "eyepiece position." It is only natural for the lay person to think that it is easy to photograph stars in this manner, especially since beautiful astrophotographs are often shown enticingly as part of promotions which emphasize *that the telescope has a motor drive.*

While one can take pictures of brightly lit distant sailboats on a sunny day and even record the full moon in this telescopic manner, it is impossible for beginners to take successful pictures of stars—let alone comets—*through the telescope* at first. The reason why it is very difficult to shoot quickly through telescopic systems at the beginning is "light loss."

When dim starlight has to move through the optical system of even a costly telescope, only a fraction of the light entering the front actually comes out the back to register on the film. The highly sensitive human eye can easily see stars in the "eyepiece position," but cameras need time. Even though such telescopes have motor drives and can arrest the motions of the stars, unreasonably long exposures are required to take such telescopic pictures. In any case, such photographs would show only tiny fields in the sky, whereas we are interested in larger areas to find the comet by our comparison method.

What we want is the *drive.* Whether it is provided by a tiny motor as in STELAS or by an actual telescope system does not matter. We want to move our *camera* with its standard lens, which loses very little starlight while collecting it.

When we mount a camera piggyback on top of an unmotorized telescope, it is even possible to "drive" telescope and camera by hand. To achieve a stop-the-earth effect, some such telescopes have linkages that allow accurate tracking of the apparent motion of the sky by simply keeping any bright star centered in the eyepiece and slowly turning a flexible shaft to make sure the star stays in the middle of the eyepiece during the length of an exposure.

In order for us to drive telescopes and cameras easily and accurately, whether by hand or with a motor, it is important that the axis on which our instrument turns is parallel to the axis of the earth. An ordinary ¼-inch plastic drinking straw, when fastened parallel to the axis of our camera or telescope and aligned with Polaris, the North Star, will give us adequate results for observations or photographs of up to 5 minutes in duration.

Do not forget: for all our Halley's Comet observations and photographs, the North Star should be directly behind us (and our cameras) or behind our left or right shoulders. Thus the camera axis (and the drinking straw) will be aimed about 40 degrees above the horizon where our little compass points north.

Owners of telescopes will do well to refer back to their telescope manual to learn simple "polar alignment" of their instruments. As for STELAS, all that you have to remember is that once you have sighted the North Star in the little paper tube, your camera system is accurately aligned with the axis of the earth. From that point forward make all your changes aiming and choosing the angles of your camera with the "ball head" (item J).

It has been said that the difference between men and boys is the cost of their toys; but the real challenge is to try to achieve the maximum results with the minimum outlay. In this connection it should be stressed here most emphatically that the simple STELAS drive system recording the heavens *under a dark sky* will outperform a camera mounted piggyback on the

costliest of telescopes under less favorable brighter skies. Under identical skies both methods and devices can discover the comet equally quickly. Ownership of a telescope is not necessary to participate in the comet rediscovery program.

Once you have found how much there is to observe, to learn and to do, or have grasped the contributions amateurs can make to science, then astronomy and astrophotography may turn into a passion. The breathtaking vastness of the universe that you can capture with a standard camera will astound you.

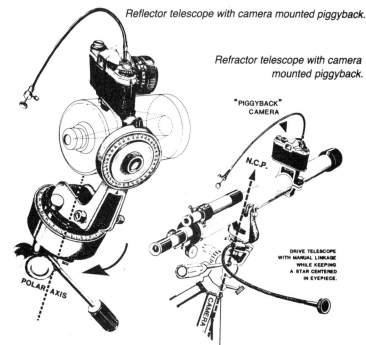

Reflector telescope with camera mounted piggyback.

Refractor telescope with camera mounted piggyback.

"PIGGYBACK" CAMERA

N.C.P.

DRIVE TELESCOPE WITH MANUAL LINKAGE WHILE KEEPING A STAR CENTERED IN EYEPIECE.

POLAR AXIS

CAMERA

13. STEBLICOM:
The Discovery Box

Comet Halley presents a target of rare good fortune to amateur and professional astronomer alike. The wealth of information gathered during previous passages allows us to know in advance exactly where it will be and when. Such a predictable object, in places known well in advance, at dates and times predetermined to the second, offers ideal opportunities. Even as you look for it you know exactly where and when to look, and you can perfect your search methods during the quest. You will survey the constellation Taurus for the comet in November 1985, scan Aries next, and concentrate on the clearly defined Halley path in Pisces during December. What we will be seeking will first register on our photographs as a small dot—possibly a fuzzy-appearing dot, not unlike the stellar images in the rest of the photographs.

As explained in Chapter 6, a foolproof method can help us to make an early rediscovery of the solar wanderer. If we just photograph the *areas* in which we know the comet to be moving at any particular time and compare such photographs with earlier ones, the comet will reveal itself. It is as if we looked at two pictures of identical regions of the sky. The first one shows a thousand stars, the second image contains a thousand and one. But we need *not* check each individual star-dot first on one photograph to see whether it also appears on the second. This would be tedious and would take a long time. We can build a simple optical comparator, a discovery box or even a machine that will help our eyes while comparing the *entire* starfield in the first picture with the same region in the second photograph. This is called blinking. It will make the comet reveal itself. (See figure at right for a graphic demonstration of the blinking principle.)

All that we need to do is examine the seemingly identical starfields in alternating fashion. Yes, the two slides will *seem* alike at first glance. Still, we *know* that the picture we took of Aquarius that warm September midnight in 1985 did not contain the comet, while our January 1985 photograph *must* show it at a magnitude well within the reach of our cameras, even from a city location.

It is important that the slides to be compared be otherwise as identical as we can make them. They should be taken from the same location, on the same type of film, *with the camera at the same angle*. Exposure times and development should be identical also. Then the comparison of the images becomes easy and fun. We put the pair of slides under the matching magnifiers in our STEBLICOM (STEreo BLInk COMparator, see page 72). Flipping one light switch on and off will instantly reveal the comet. It will not fail to fascinate even those professing disinterest in astronomical science. You will be featuring the most famous comet in history.

In the double night lights of a STEBLICOM it will become clear why it can be rewarding to practice one's photography early during 1985 to try for the best results. From a rooftop, a garden or even a park, a camera mounted on a stop-the-earth motor drive assumes remarkable capabilities. Our discovery

The Blinking Principle: To obtain a "blinking" effect, flip this corner starfield rapidly back and forth over the one below and discover a comet.

Reference slide: Taken two nights ago, last week or even last month.

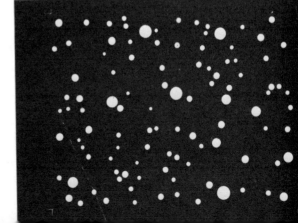

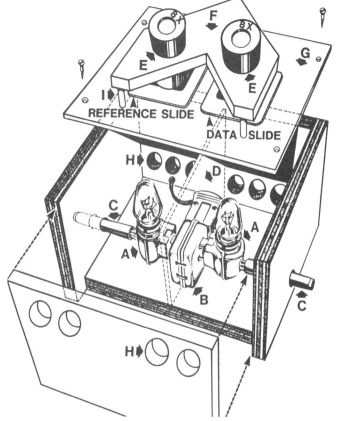

A▶ TWO 7 WATT NIGHT LIGHTS
WITH ON/OFF SWITCHES.

B▶ THREE-WAY PLUG
EPOXY GLUE TO BASE-PLATE.

C▶ 3/8" PLASTIC TUBING
OR USE FELT-PEN CAPS.

D▶ LAMP DIVIDER SHIELD
4"X5" LIGHTPROOF CARDBOARD.

E▶ TWO 6-8 POWER MAGNIFIERS
OR COMPARATORS WITHOUT RETICLES.

F▶ MAGNIFIER YOKE SUPPORT
MEASURE DISTANCE BETWEEN PUPILS.

G▶ 7"X6" WHITE ACRYLIC TOP
OR USE FROSTED PLEXIGLASS.

H▶ VENTILATION HOLES
CAN BE TRIANGULAR CUTOUTS.

I▶ (3)WOODEN SPACER DOWELS
TO PERMIT ALIGNMENT OF SLIDES.

This is a home-built STEBLICOM. Any sturdy box will serve as a housing for the night lights and will support a frosted sheet of acrylic. You need switch only one of the two lamps and can do so with any simple breaker. Use only UL-approved components for safety.

boxes or machines will place 1985–86 cometwatchers where even scientists were unable to reach in 1910.

Two discovery modes present themselves to the modern comparison blinker. When we compare a September 1985 photo of Pisces with a seemingly identical December photograph, we have a no/yes situation. We know in advance that the comet will *not* be in the earlier photograph. It will, however, be in the later image, provided we photographed stars as faint as magnitude 7. (The earlier photograph in blinking is always referred to as the "reference" slide, while later pictures are "data" slides (see figures in lower right pages 71 and 73).

In the case of taking a matching pair of photographs, say of the constellation Aquarius one night apart at the beginning of January 1986, a yes/yes situation will occur. If we have recorded stars to 6th magnitude (easy!) the comet will show in both the reference and the data slides. Instead of blinking on and off as in the earlier pair, the comet will jump back and forth along the line of its path. In either case the comet will quickly attract attention to itself, which is why both kinds of blinking pairs can be employed for early rediscovery.

There are many advantages in starting to take photographs of the various host constellations as long in advance as possible. Not only will you be gaining important practice, but you will be building a small selection of reference slides from which you can later select one that best duplicates a particular data slide for easiest blink comparison. Please *never ever* throw a single slide away. Keep each one with all the notations on it. Here is why: Just as in the real scientific world, once you have found the comet and know what to look for, you will reexamine all your earlier slides only to find a "prediscovery" photograph that may predate your "discovery" photograph by four weeks. Invariably the object of your search will have turned up in the slide that had the streaks of airplane lights running through the bottom from edge to edge, the one slide you were going to dump.

In the case of STEBLICOM blinking it is possible to rotate both slides as needed under the magnifiers to correct for slight variations in the angle at which the photographs were taken.

Professional astronomers use blink comparators all the time. Their sophisticated systems have price tags in the range of five or six figures. But they are invaluable tools, and many discoveries are being made with them. Even if you blink only limited numbers of pairs of photos taken days, weeks or even months apart, you may record some of the thousands of variable stars that will appear faint in one slide and have brightened surprisingly in another. Amateurs play an important role in variable-star research, and you may wish to learn more about it. Contact the American Association of Variable Star Observers (AAVSO) and send them a self-addressed double-stamped envelope. In time you will receive information on challenging science opportunities in which you can participate.

Quite serendipitously you may discover a precious nova or a comet in one of your photographs. Even without such a stroke of luck you will quickly recognize what powerful scientific tools the combination of photography with blinking has placed into your hands. You may soon wish to set out on your own systematic astronomical search program (see Chapter 14).

For a simple demonstration of how STEBLICOM works, rapidly flip the starfield on page 71 back and forth over the one shown here. "Blinking" will soon explain itself.

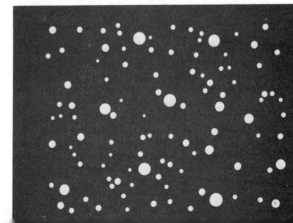

Data slide: Your most recent slide of the same area as on page 71.

Comets in morning haystack.

14. Comets in a Haystack:

They are really formed like haystacks, those morning and evening cone-shaped regions in which new comets can be found. They surround the ecliptic, the imaginary line along which the sun seems to travel through the year. As we know, the sun only *appears* to be journeying along this illusory path from the time we first spy it on a sparkling morning until it disappears in

How to Discover Your Own Comet

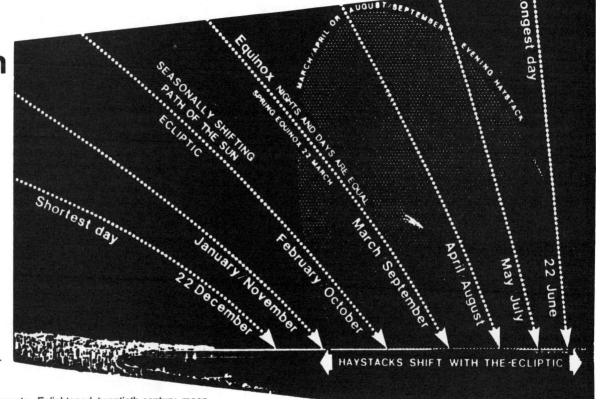

SEASONALLY SHIFTING PATH OF THE SUN

ECLIPTIC

Equinox

NIGHTS AND DAYS ARE EQUAL

SPRING EQUINOX 22 MARCH

MARCH/APRIL OR AUGUST/SEPTEMBER

EVENING HAYSTACK

Longest day

Shortest day

22 December

January/November

February/October

March/September

April/August

May/July

22 June

HAYSTACKS SHIFT WITH THE ECLIPTIC

Comets in evening haystack.

the blaze of magnificent sunsets. Enlightened twentieth-century moon-walkers that we are, we know today that the sun, our nearest star, stands immobile in the heavens. It is the rotation of our planet that creates the illusion of solar motion and the appearance that all about us rises in the east and sets in the west.

A planet called Earth turning on its axis, imagined motions of suns standing still in the heavens, fanciful haystack-shaped regions with journeying comets? It may sound like a scenario out of *Alice in Wonderland*, and in a way it is. As for trying to find a comet in the morning or evening skies, it used to be a little like finding a needle in a haystack, but no more.

Cometwatchers still hold long predawn vigils and spend many hours after sunset visually scanning the eastern and western skies for the first glimpse of a comet never seen before. As in our own observations of Halley's Comet, we know that cometary visitors are best seen going into their turn around the sun or coming out of it. Even though comets may be discovered anywhere in the heavens, chances for finding them are improved in the general region of the haystacks (see pages 74 and 75).

Here, then, are promising areas in which to put our cameras to work. We can regularly collect morning and/or evening photographs of the haystack areas, allowing the faint light of a tenuous comet to collect on the emulsion of our film. Instead of our spending hours searching the skies at the eyepiece of a telescope, a few carefully aimed photos can capture the first moments of darkness almost anywhere above the sunset skyline or the last minutes just above the horizons of dawn. When we develop and then blink such pictures a new and reliable method for comet discoveries presents itself.

It stands to reason that for such comet patrol work good dark unimpaired views of the eastern and western horizons are best. Truly dedicated telescopic comet seekers (and they are few in number) sometimes travel to points east of their cities for morning telescopic patrols and west to set up their telescopes for evening scanning.

In Adelaide, Australia, William Bradfield, a truly committed amateur in search of comets, has twelve discoveries to his name at this writing. In Japan, Minoru Honda has made multiple discoveries both of comets and of novae. Recently Leslie Peltier held the record for comet discoveries in the Americas: he also had twelve.

You do not have to make multiple discoveries: a single find will bring the network television crews to your door. Since you will be working with reference and data slides, you should always go back the morning or evening after your first discovery and take a confirming photograph or two. The object must have moved, just as you will see Halley's image moving from day to day. You may tremble in anticipation as you develop the black-and-white film, because if the object is still there twenty-four hours later, shifted in any direction at all, you may have found a comet. Check with your nearest college to be sure you have not photographed the planet Uranus (planets also move about in the heavens). After that you are ready to send a telegram to the Smithsonian Astrophysical Observatory (TELEX code: TWX 710-320-6842, ASTROGRAM CAM.), in which you should stress *that you have before and after discovery photographs.* In the meantime get some prints made (you will have blinked your negatives, of course) and write a follow-up letter into which you copy *all* the notes from your log. The data about your camera, lens, film, exposure times, location, dates, times of day or night, etc., all become critically important now. Enclose your reference and data prints (keep the masters safe). Don't forget to give your home phone number. Send your follow-up letter by overnight express to:

The Central Bureau for Astronomical Telegrams, Smithsonian Astrophysical Observatory, Cambridge, MA 02138, U.S.A.

Stay by the telephone. If you have found a comet it will surely ring.

RESOURCES

Bibliography and Recommended Reading

Edberg, Stephen J., *International Halley Watch Amateur Observers' Manual for Scientific Comet Studies,* 2 parts, published by the National Aeronautical and Space Administration. Washington: U.S. Government Printing Office, 1983.

Mayer, Ben, *Starwatch.* New York: The Putnam Publishing Group, 1984.

Menzel, D. H., and J. M. Pasachoff, *A Field Guide to the Stars and Planets,* 2nd edition. Boston: Houghton Mifflin Company, 1983.

Vehrenberg, Hans, *Atlas of Deep Sky Splendors.* Cambridge, MA: Sky Publishing Corporation, 1978.

Telescopes and Related Equipment

Bushnell/Bausch & Lomb
2828 East Foothill Boulevard, Pasadena, CA 91107

Celestron International
2835 Columbia Street, Torrance, CA 90503

Coulter Optical
P.O. Box K, Idyllwild, CA 92349

Edmund Scientific
101 East Gloucester Pike, Barrington, NJ 07092

Meade Instruments
1675 Toronto Way, Costa Mesa, CA 92626

Orion Telescopes
P.O. Box 1158-T, Santa Cruz, CA 95061

Questar Corporation
P.O. Box C, New Hope, PA 18938

Roger W. Tuthill
11 Tanglewood Lane, Mountainside, NJ 07092

Magazines

Astronomy/Deep Sky
Astromedia Corporation
P.O. Box 92788, Milwaukee, WI 53202

Griffith Observer
Griffith Observatory
2800 East Observatory Road, Los Angeles, CA 90027

Mercury
Astronomical Society of the Pacific
1290 24th Avenue, San Francisco, CA 94122

Sky and Telescope
Sky Publishing Corporation
49 Bay State Road, Cambridge, MA 02238

Telescope Making
Astromedia Corporation
P.O. Box 92778, Milwaukee, WI 53202

Star Atlases

Norton's Star Atlas
Sky Publishing Corporation
49 Bay State Road, Cambridge, MA 02238

Tirion Atlas 2000.0
Sky Publishing Corporation
49 Bay State Road, Cambridge, MA 02238

AAVSO Variable Star Atlas
American Association of Variable Star Observers
187 Concord Avenue, Cambridge MA 02138

Posters, Slides, Photographs

Hansen Planetarium Publications
15 South State Street, Salt Lake City, UT 84111

Sky Calendar (Monthly)

Abrams Planetarium
Michigan State University, East Lansing, MI 48824

STELAS and STEBLICOM

Celestron International
2835 Columbia Street, Torrance, CA 90503

GLOSSARY

APERTURE. Effective light-gathering diameter, such as the opening of a lens.

ASTEROID. A minor planet.

CELESTIAL EQUATOR. A great circle on the celestial sphere beyond and in the same plane as the equator of the earth.

COMET. A body of ices, rock and dusty matter.

CONSTELLATION. A grouping of stars named for mythical figures, animals or objects.

CULMINATION. The point of highest altitude in the sky of a star or a constellation.

DAYLIGHT SAVING TIME. Time more advanced by one hour than Standard Time.

ECLIPTIC. The apparent sky path of the sun.

EQUATORIAL DRIVE. A motorized mounting with an axis parallel to the earth's axis whose motion compensates for the rotation of our planet.

GALAXY. A vast assemblage of millions to hundreds of thousands of millions of stars.

HAYSTACK. Horizon zones of haystack shape centered on the ecliptic, where comets can be found before sunrise and/or after sunset.

LATITUDE. A north–south coordinate on the surface of the earth.

LIGHT-YEAR. A measure of distance, not time. The distance that light travels in one year.

MAGNITUDE. A scale of measurement for the brightness of stars and other sky objects.

METEOR. A body of rock or metal heated to incandescence when it enters the atmosphere of the earth. Also called a ''shooting star.''

METEORITE. Surviving part of a meteor which strikes the earth after a fiery descent.

METEOR SHOWER. A display of many meteors radiating from a common point in the sky.

MILKY WAY. A wide band of light stretching around the celestial sphere caused by the light of myriads of faint stars.

NOVA. A star that suddenly undergoes an outburst of radiant energy and increases its luminosity by hundreds or thousands of times.

ORBIT. The path of a body in its revolution about another body or center of gravity.

PLANET. One of nine large spherical objects revolving about the sun and shining by reflected light.

PROBLICOM. A PROjection BLInk COMparator to present two different photographs of the same region of the sky for easy comparison.

RADIANT. The point on the celestial sphere from which meteors of a given shower seem to originate.

SCALE. The linear distance in a photograph corresponding to a particular angular distance in the sky—i.e., so many fractions of an inch per arc-minute or per degree.

SERENDIPITY. The gift of finding valuable or useful things or data not sought for.

SPORADIC METEOR. A meteor (shooting star) that does not belong to a meteor shower.

STARFRAME. A sky area defined by a specific outline (such as a bent coathanger) to contain the principal area of a constellation for easy repeated finding and study.

STAR PARTY. A gathering of amateur starlovers on new-moon weekend nights to view the skies, take photographs, compare equipment and share astronomical information.

STEBLICOM. A STereo BLInk COMparator for viewing two different photographs of the same region of the sky for easy comparison.

SUPERNOVA. A stellar cataclysm in which a star explodes, briefly increasing its luminosity by hundreds of thousands to hundreds of millions of times.

VARIABLE STAR. A star whose brightness varies.

ZENITH. The point directly overhead, the direction opposite to that of a plumb bob.

ZODIAC. An imaginary band around the sky which has the ecliptic in its center.